'An extraordinary book, the very best of science writing, because it is not just about science, but is memoir, history, bleeding-edge genetics and a completely original take on original sin. Thrilling, entertaining, provocative, brilliant'
Adam Rutherford, author of
A Brief History of Everyone Who Ever Lived

'A powerful read that stops you dead in your tracks and forces you to think very deeply'
Professor Sue Black, author of *All That Remains*

'A tour de force . . . a thoughtful and thought-provoking book that invites us to go deep into questions about why people do terrible things and how we should treat them afterwards. Harden's discussion is deepened by her personal reflections on her own responses to hurt and cruelty – a rare mixture, showing how the scientific and the personal perspective combine in a rich complementarity. I loved this book'
Gwen Adshead, author of *The Devil You Know*

'Regardless of the side you take in the nature vs nurture debate, *Original Sin* offers an eye-opening perspective'
Patric Gagne, author of *Sociopath*

'Unique, expansive and illuminating – a mix of religion and genetics that interweaves intensely personal storytelling with rigidly objective science to explore big questions about the bad things we have the capacity to do'
John Higgs, author of *William Blake vs the World*

'Even if you have no interest in the concept of sin, this is a compelling read. Harden's bottom line is that we subjective beings are morally responsible for our actions (sorry), despite the fact that we also tick along deterministically. This emotionally startling and intellectually erudite book explains why'
Mark Solms, neuroscientist, psychoanalyst and author of
The Only Cure

BY KATHRYN PAIGE HARDEN

The Genetic Lottery

Original Sin

ORIGINAL SIN

ORIGINAL SIN

The Genetics of Wrongdoing, the Problem of Blame
and the Future of Forgiveness

Kathryn Paige Harden

First published in the United States in 2026 by Random House
This edition first published in Great Britain in 2026 by Weidenfeld & Nicolson,
an imprint of The Orion Publishing Group Ltd
Carmelite House, 50 Victoria Embankment
London EC4Y 0DZ

An Hachette UK Company

The authorised representative in the EEA is Hachette Ireland,
8 Castlecourt Centre, Dublin 15, D15 XTP3, Ireland
(email: info@hbgi.ie)

1 3 5 7 9 10 8 6 4 2

Copyright © Sophrosyne Studio LLC 2026

The moral right of Kathryn Paige Harden to be identified as
the author of this work has been asserted in accordance
with the Copyright, Designs and Patents Act of 1988.

All rights reserved. No part of this publication may be
reproduced, stored in a retrieval system, or transmitted
in any form or by any means, electronic, mechanical,
photocopying, recording, or otherwise, without the
prior permission of both the copyright owner and the
above publisher of this book.

A CIP catalogue record for this book is
available from the British Library.

Grateful acknowledgment is made to the following for permission
to reprint previously published material:
Cambridge University Press c/o Copyright Clearance Center: Excerpt from
'Cain (A New Play)' by Howard Nemerov (*Tulane Drama Review*,
vol. 4, issue 2, December, 1959). Reprinted by permission.
HarperCollins Publishers and The Estate of Louise Glück c/o
The Wylie Agency LLC: 'Clover' from *The Wild Iris* by Louise Glück,
copyright © 1992 by Louise Glück. After May 1, 2027, rights to this poem
are controlled by The Estate of Louise Glück c/o The Wylie Agency.
Reprinted by permission.
The MIT Press: Excerpt from *Contours of Agency: Essays on Themes from Harry
Frankfurt* edited by Sarah Buss and Lee Overton, copyright © 2002 by
Massachusetts Institute of Technology. Reprinted by permission.
Lauren Slater c/o *The Missouri Review*: Excerpt from 'Black Swans' by Lauren Slater
(*The Missouri Review*, Spring, 1996.) Reprinted by permission.

ISBN (Hardback) 978 1 3996 0433 8
ISBN (Export Trade Paperback) 978 1 3996 0434 5
ISBN (Ebook) 978 1 3996 0436 9
ISBN (Audio) 978 1 3996 0437 6

Typeset by Input Data Services Ltd, Bridgwater, Somerset

Printed in Great Britain by Clays Ltd, Elcograf, S.p.A.

for Travis

Let the reader, where we are equally confident, stride on
    with me;
where we are equally puzzled, pause to investigate with me;
where he finds himself in error, come to my side;
where he finds me erring, call me to his side.

St. Augustine

Without contraries is no progression.

William Blake

Contents

Desert	3
Sin	13
Letter	35
Luck	36
Animal	61
Choice	86
Essence	112
Variety	138
Consequences	165
Retribution	190
Eclipse	216
Puzzle	239
Acknowledgments	251
Notes	253
Index	293

ORIGINAL SIN

Desert

A prologue, in which I take a psychedelic drug and realize how easily the illusion of self can dissolve, even as the feeling of responsibility lingers

IT WAS A SWELTERING, tedious summer, the first summer of the pandemic, when Travis and I decided to go to the West Texas desert to drop acid. Our itinerary did, of course, include things other than taking drugs. We plan a day hiking in Big Bend. We book a Donald Judd–esque gray concrete casita with a view of the Chisos Mountains. We look forward to drinking Modelo and doing crossword puzzles and having morning sex. But lovely as all of that will be, our main purpose is to take LSD.

A friend in Berlin mails me a few tabs taped under a handmade construction-paper collage on the front of a fake birthday card, posted with a bogus return address. The post office has been running slow all summer, the latest outrage from the first Trump administration, and I am relieved when the card arrives in time for our departure.

Travis and I have been trapped in the house with my children for months. They were still my children then, the fruits of my first marriage, but they are slowly becoming our children, as Travis works to be more to them than just their mother's boyfriend. He holds them, emotionally and physically, as they rage against their quarantine walls. Watching him earn their trust, I realize how close you can get to a wild animal simply by being very still.

But stillness is taxing. Our lives have become a hot and dreary cycle of Sisyphean sameness. We want to escape, not just our routines but ourselves. We want life to be different, and life cannot be different. Trump is still president; the virus still spreads. The children and the dog still insist on being fed every day. So Travis and I decide to make our brains different, a hard serotonergic reboot, courtesy of international airmail. My ex-husband agrees to take the children for a long weekend, so that I can get away. "You're going to do what?!," he exclaims down the phone, and we are both glad, once again, that we are no longer each other's keeper.

Before this trip to West Texas, I had used LSD only once before, just a few weeks after Travis and I met. In the fall of 2019, I was hurting from a brief, intense, spoiled love affair with a swaggering artist. I was eager to fall in love, wary of falling in love, sometimes within the same five minutes. Michael Pollan's book on LSD convinced me that what I really needed to sort myself out was an "experience of the numinous." A friend, the same one who would later mail me the fake birthday card, rented a house in Colorado, set on nine acres full of yellow aspen trees, and we arranged for two more friends to come have a psychedelic experience with us. The ancient Greeks sought to understand the numen through watching the motions of birds; we were going to take a lab-synthesized molecule and watch the leaves quiver. The night before I left for Colorado, Travis insisted we should wait to have sex for the first time until after I came back from my trip. "You'll come back with your third eye, and then we'll fuck," he said. "You'll be glad we waited."

After my first trip, I texted him three things. First, a message that I had had "the most perfect, spectacular day." Second, a picture of me in a field of tall golden grass, my face filled with surprise and delight. And, third, a passage from an essay by psychologist and writer Lauren Slater:

> We have come to think, lately, of machines and animals, of machines and nature, as occupying opposite sides of the spectrum—there is IBM and then there's the lake—but really they are so similar. A computer goes on when you push its button. A gazelle goes on when it sees a lynx. Only humans are supposedly different, above the pure cause and effect of the hard-wired primitive world. Free will and all.
>
> But no, maybe not. For I had swallowed a pill designed through technology, and in doing so, I was discovering myself embedded in an animal world. I was a purely chemical being, mood and personality seeping through serotonin. We are all taught to believe it's true. But how strange to feel that supposed truth bubbling right in your own tweaked brain pan. Who was I, all skin and worm; all herd?

Slater has OCD, and her essay is about Prozac, another drug that changes serotonin. Prozac binds to serotonin receptors on "pre-synaptic" neurons—brain cells that are sending out serotonin as a chemical messenger—blocking them from reabsorbing serotonin too quickly. Instead of binding to pre-synaptic receptors, LSD binds to post-synaptic ones. Both drugs affect us because they are very similar to a molecule we already produce in our brain. And by changing the brain's chemical messaging, both drugs can cause lasting changes in mood, personality, and sense of self, effects that many scientists think are due to the serotonin system's effects on neural plasticity. The brain on

Prozac or LSD is forging new trails of association and thought, instead of being confined to the same ruts of rumination and habit.

I had assigned Slater's essay to my undergraduates before. Like her, I'm a psychologist, and I've taught thousands of psychology majors about serotonin, synapses, and the neurobiology of mental illnesses like OCD. But after my first acid trip, I finally believed: I am a chemical being. I am embedded in an animal world. Experiencing is a different form of understanding.

Coming home to Austin from Colorado, that feeling of being all skin and worm, all herd, lingered through airport security and the crowded plane, through the school carpool line and the children's bedtime routine. I felt like I had sloughed off a callused psychological skin, and now light and touch and sound were achingly vivid.

When Travis arrived that night, announcing himself with a text saying "here" rather than knock and potentially wake up the children, I was indeed glad we waited. The best creative acts feel transcendent. Awareness of space and time recedes and with them your sense of boundedness. Whose limbs are these? Whose tongue is whose? Self-consciousness abates, as does distraction, and you find yourself instinctively doing, without much thinking. The gazelle turns on. I had taken a drug designed through technology and learned how flimsily constructed my ordinary self-consciousness was. So it was easier, now, to let go of my illusions of separateness, my grasps at control. I could observe, without embarrassment, how I was participating in a primal chain of cause and effect. Pull, push. Tug, gasp. Enter, release.

Do you see, then, why we thought that a psychedelic drug would be a good way to counter the helplessness of our pandemic summer?

·········

The day before we take LSD together, Travis and I travel to a five-acre patch of desert, forsaken by gods and humans alike, outside Terlingua. Population: 58. He owns this bit of land, but he has never seen it before. Or rather, he co-owns it, with three other architects. They purchased it before beginning graduate school at the University of Texas, because if you own land in Texas then you qualify for in-state tuition. Many years later, Travis still suffers under a staggering burden of student debt, but his debts are smaller than they might have been if he hadn't bought a tiny piece of hell.

We ease my soccer-mom SUV over dusty tracks made by mining equipment until we can't go any farther. We get out and walk. We scan the ground for fossils. This arid land used to be at the bottom of an ancient inland sea.

I am fine, and then suddenly I am not. I become disoriented and faint. My vision blurs; I can't think straight. I return to the car, blast the air conditioner, sip from a bottle of water, try not to think about how desperate we would be if the car died and we had to hike our way out of there. I try not to think about how deserts can kill people.

Marcia Powell died in the desert. She was forty-eight years old when she was placed in an outdoor, uncovered holding cell in a prison in Arizona. The temperature that day reached 107 degrees. Prisoners were supposed to be held in such holding cells for "only" two hours. She was found unresponsive in her cell, after being there for nearly four hours, with burn blisters on her arms, legs, chest, and face. She died in the local hospital that night. Cause of death: "Complications of Hyperthermia due to Environmental Heat Exposure." Three guards, who had been stationed twenty yards away from Powell's cage, were suspended while her death was investigated. But ultimately her death was declared an "accident." No one in the Arizona Department of Corrections was held legally accountable.

Powell was serving a twenty-seven-month prison sentence for offering a blowjob to a police officer. Powell was doing sex work to get money for drugs. She didn't get her drugs conveniently mailed to her house from overseas, and she wasn't using them to dissipate her bourgeois ennui. She had, according to court reports, a serious polysubstance addiction that was complicated by untreated disorganized schizophrenia and mild intellectual disability. Like the damned souls in the Seventh Circle of Hell, Powell was burned in "an expanse of deep and arid sand."

Philosophers use the word *desert* to refer not to an arid landscape but to the condition of deserving something by virtue of one's behavior or conduct. What you deserve are your "just deserts." A child deserves chocolate cake because he has cleaned up his room. A mentally ill woman deserves to be caged in the desert, because she offered to put a man's cock in her mouth for money and drugs. Or so says the state of Arizona.

Our car did not get stuck. I recovered from my mild heat exhaustion. We returned to the hotel, happy and hungry, but with a heightened vigilance to the desert's dangers. The next day, Travis and I prepared snacks and water and then dissolved the acid tabs under our tongues. We gently took off each other's clothes and climbed into bed.

My second trip begins as a tunnel of heightened internal sensations: an awareness of my throat as I swallow, the texture of the sheets against my skin, the tiny ripple of contractions in my legs as I move. Then I am mesmerized by Travis's face, which reminds me of a Dürer self-portrait even when I'm stone-cold sober and which looks even more handsome and grave and beatific while I am tripping. He kneels at the end of the bed, framed by the casita windows, austere cut-outs in the gray concrete, the ocotillos burning fluorescent green behind him. "I will never recover from the way you look right now," I tell him—and I haven't, still.

He is, in ordinary life, emotionally reserved. Over time, I've learned that his reserve is not callousness or unaffectedness, but rather the conditioned ability to mask what are, in fact, deep wells of feeling. While on acid, however, that stoic mask dissolves. He is laughing with delight, until he isn't. I emerge from an internal reverie to find him in the shower, crumpled on the floor, convinced something terrible has happened.

For the next few hours, as my own mental state returns to normal, he continues to slip in and out of intense paranoia. In defiance of social distancing rules, a rich girl and her friends are renting out the rest of the hotel, celebrating her twenty-fifth birthday party with oysters they have flown in from Massachusetts. The sounds of their gaiety periodically lull him into a feeling of safety, and then he is tugged under another wave of hallucination and derealization.

Later, much later, I was finally able to piece together what his mind created while tripping: He had caused a car wreck, and I was dead, and he was to blame. His crumpled posture, the strange angle of his arm: He was bleeding out on the pavement. There was blood on his hands. He had caused the car wreck, and I was dead, and he was to blame. He kept uncurling his body and standing up, trying to get back to the point where he could avoid the coming tragedy, trying to rewind time, like Superman flying the wrong way around the globe, but to no avail. The scene would replay itself in the exact same way every time, an inexorable loop. In his mind, he could observe himself participating in a chain of causes and effects, and the effects were awful. But his horror and guilt about how things turned out did not give him the power to change course. He had caused the car wreck, and I was dead, and he was to blame.

We had hoped to escape ourselves. But for him, the drugs instead summoned a daytime nightmare, both anguished and accusing. He did escape the ordinary feeling of an agentic, bounded self, an ego that feels separated from the pure cause and effect of the primitive world. But even as his sense of self-as-causal-agent disappeared, he was overwhelmed by feelings of responsibility for the consequences. He was absolutely convinced of his blameworthiness; he was convicted by it. And he was helpless in the face of that conviction; there was nothing he could do. There was nothing he could have done. He could not have done otherwise. Nonetheless, he had blood on his hands.

It was only a dream. Or, rather, it was only a hallucination, a type of waking dream, and it wasn't even mine. But the experience of watching my lover experience visible torment, terror, and guilt lingered in my mind, long after the LSD molecules finally came unbound from Travis's serotonin receptors, long after he had finished processing his thoughts and feelings with me and other trusted confidants. Like a biblical parable or Zen koan or Freudian slip, the trip seemed to both invite and evade exposition. It wasn't my trip, but it became my dream interpretation project anyway.

The psychoanalytic view of dreams and nightmares is that they are symbolically rich narratives that reveal our unconscious wishes, desires, fantasies, or fears. They tell us about what is already in our minds. They are, as Freud put it, in "that class of the terrifying which leads back to something long known to us, once very familiar." The worst nightmares whisper dark truths that you've been unable to speak out loud, that you've barely been able to admit to yourself, but that you immediately recognize, old acquaintances of your heart: You hate her. You want to fuck him. You failed. People are laughing at you. People can see what you've tried to hide. Your child is going to die one day. You are going to die one day. You did it. You are to blame. Nightmares uncover beliefs we have locked away in our secret rooms, repressed but never fully forgotten.

What was the ghastly belief at the core of Travis's waking nightmare? "There's nothing you can do about it, and you will be responsible." This idea runs counter to the story we like to tell ourselves about blame and responsibility. We like to believe that we will be blamed for our choices, that we won't be found guilty unless we have had the opportunity to make different choices, that we can avoid damnation by making better choices. We won't wriggle on the hook unless we have had a chance not to take the bait.

The philosopher Friedrich Nietzsche suggested that this neat story about blame and punishment is new: "That thought—'the miscreant deserves punishment *because* he might have acted otherwise,' in spite of the fact that it now seems so trite, obvious, natural and inevitable . . . is in point of fact an exceedingly recent, even refined form of human judgement and inference." In the cracks in our consciousness, in our nightmares and serotonin-soaked hallucinations, we catch glimpses of an older story, one we fear might still be true. Once upon a time, before we invented villains, there were simply enemies, and we could, unwittingly, become the enemy.

We encounter this older story in the Hebrew Bible. "Suppose you sin by violating one of the Lord's commands. Even if you are unaware of what you have done, you are guilty and will be punished for your sin." We encounter it in Greek tragedy. Oedipus did not know what he was doing when he killed his father and wed his mother, but he gouged his own eyes out all the same, proclaiming, "This punishment that I have laid upon myself is just."

As I came to realize in the aftermath of our trip, and as I will describe in the pages to come, ancient texts and altered states are not the only places we are reminded of the complicated relationship between control and blame. We also hear whispers of this older story in modern science. My beloved's hallucination haunted me, because it raised questions already burbling at the edges of my consciousness, already raised by my work as a scientist. Questions about agency, responsibility, transgression, punishment, and having a meatcomputer for a mind.

When we returned to Austin, I began writing in my journal, trying to sort out my thoughts. I began reading philosophy, theology, and fiction that considered the same dark themes as an LSD-soaked brain. The more I read, the more questions I had; the more questions I had, the more I wrote.

The result is this book.

Sin

In which I consider the ancient question of whether our biological inheritance diminishes our human blameworthiness and struggle to find certainty

IN THE FALL OF 2021, I received a letter in my university mailbox from a man imprisoned in the J. Dale Wainwright Unit, formerly known as the Eastham Unit, one of the oldest prisons in Texas. The prison is isolated, rural, enormous—a "God-forsaken hole." Clyde Barrow, of Bonnie and Clyde, was its most famous detainee. The land occupied by the prison complex was originally cleared by enslaved men and women. After the Civil War, the family after whom the prison was initially named, the Easthams, ran their farm by leasing convict labor from the state of Texas. Even today, the Wainwright Unit remains a large-scale agricultural operation, with cattle, hogs, laying hens, and hundreds of acres of crops, all powered by unfree labor. The man wrote in his letter that he had been in Wainwright since he was sixteen years old.

Several cartoons and illustrations were taped to the letter, and I couldn't help but laugh at the dark wit that selected them. In one, a bearded caveman lies on a couch, talking to the psychoanalyst behind him: "Now that I have a prefrontal cortex I worry about everything." In another, a white tiger on the psychoanalyst's couch says, "They train me to perform, then when I try to show off what I really do best, everybody goes ballistic." The third was an abstract graphic, a cubist-esque collage of facial features overlaid with dots and dashes and bar codes that suggested something about humanity being atomized, automatized, analyzed. It was from an article in *Texas Monthly* magazine about me.

The article described the research we conduct in the Developmental Behavior Genetics lab that I run as a professor of clinical psychology at the University of Texas. "Developmental": We primarily study children and adolescents, who are still growing and changing. "Genetics": We study people's DNA sequences (their genomes), as well as chemical tags attached to their DNA that affect how their genomes work (their epigenomes). "Behavior": As psychologists, we are interested in how people's genes combine with their environments to affect their thoughts, feelings, and actions in the world. The article quoted critics who consider the study of behavioral genetics to be "dangerous nonsense," but my correspondent's interest in my area of research was undeterred. He wrote that he agreed with the premise of our work. He thought that genetics had insights about human behavior that many people chose to ignore. He had ten questions he hoped I could answer. One made me catch my breath:

> Why would a young boy of 16 attack a total stranger, a female, at knife point in broad daylight at a busy intersection and make the female drive against her will to sexually assault her? What would drive a boy to do such a thing?

I looked him up online. My correspondent had been sentenced to thirty-five years in prison for aggravated kidnapping, aggravated robbery, and aggravated sexual assault.

His crime is unimaginable to me. As in, I literally cannot imagine committing such an atrocious, violent act. When I force myself to think about it, I can instead only feel hints of what his victim must have felt. I can readily sympathize with what it feels like to be hurt by male physical violence; I cannot readily imagine what it feels like to hurt someone like that.

His question, however, feels achingly familiar: *Why? What would drive me to do such a thing?* Who has not asked this very question about some action we have come to rue? The question "Why?" is not always a mere request for information. "Why?" can be a search for absolution, or at least for compassion. In a previous age, he might have posed his question to a priest in the confessional or saved it for his prayers to an unseen god. Only recently have scientists been expected to answer questions about why people do horrible things to each other.

That same fall, a few weeks before I received the letter, my colleagues and I published a paper in a scientific journal, *Nature Neuroscience*, that described results from a project we had been working on for nearly four years. Publishing any scientific paper is an ordeal that involves repeated caustic rejection, and this paper was no exception.

I had received news of one such early rejection the day before Travis and I left for West Texas. One of my co-authors sent a dejected email: "I'm seriously at a loss." She had just gotten the anonymous critiques back from three people the journal had asked to be reviewers. They didn't like it. One reviewer complained: "The main text goes through a seemingly endless list of prediction analyses . . . substance use/disorder, psychiatric illness, suicide, criminal convictions, diabetes, liver cirrhosis, condomless sex, and so on."

What we were using to predict this "endless" list of behaviors and addictions and medical diseases and personal tragedies and criminal records was people's DNA. We first pooled together data on the DNA of 1.5 million people, and then we analyzed it for patterns: Which DNA sequences were more common in people who smoked pot or smoked cigarettes or abused alcohol or had a lot of sex or described themselves as more impulsive and risk-taking? We then took the patterns we had discovered, applied them to DNA collected from new samples of people, and tested whether their total genetic "score" could predict their life outcomes. And we found that it could: Less than 20 percent of people with a low genetic score were ever arrested for a crime, for example, compared to nearly 40 percent of people with a high genetic score. (All the people in the study had genetics most similar to people whose recent ancestors lived in Northern Europe, and all identified, socially, as white.)

Travis calls this my *Minority Report* paper, after the Philip K. Dick novella and the Steven Spielberg movie of the same name, in which three "precogs" see into the future and predict crimes before they happen. I admit he's right that the project does seem like something from a science fiction dystopia: Here, spit into this tube, and a machine will read part of your DNA sequence from the cells in your saliva, and based on that DNA we will make a prediction about the odds of you being arrested for a crime, becoming addicted to alcohol, or using opiate drugs.

The precogs could peer into the future, but sometimes they disagreed about what was to come. Similarly, DNA cannot definitely say that you will commit a crime. We can say whether, based on your DNA, you are in a high-risk group, whose probability of being arrested for a crime is twice as high as that of people in the low-risk group—but that probability is still far from 100 percent. There's a yawning gap between being able to say that genetics makes a difference for violent crime rates and being able to say that this person *will* commit a crime because of their genes. You could find that gap frustrating or relieving.

Our reviewer didn't object to our paper on the grounds that it had *Minority Report*–esque dystopian elements. They were just frustrated by its "nonspecificity." They couldn't see the scarlet thread that connected the many behaviors that we analyzed in a single paper: crime and addiction and sex and suicide. What was the *theory*, they wanted to know, that connected these analyses?

The reviewer couldn't see: We were studying the genetics of sin. Or, rather, we were studying the genetics of behavior that Christianity calls sinful. Rather than presenting an "endless" list of genetic prediction analyses, we could have simply enumerated the circles of Hell.

Not that the terms we use in a scientific paper would necessarily be recognizable to Dante. The language of modern scientific psychology studiously avoids any whiff of moralizing. We didn't talk about "the vice of lechery"; we just presented statistics on how one's DNA is related to the "number of self-reported sexual partners." We didn't name "those who by violence do injury to others"; we added up how many symptoms of conduct disorder a person has, according to the American Psychiatric Association's *Diagnostic and Statistical Manual*.

But every single analysis in our paper could be mapped to *The Inferno*. In the Second Circle, the sexually licentious and wanton are blown around endlessly by whirling vortex winds. In the Third Circle, the gluttonous, who ate too much and drank too much and who placed their addictions above all else, slosh around in a putrid stew, clawed by the three-headed dog Cerberus. In the first part of the Seventh Circle, a scalding river of blood torments those who were violent against others; the suicidal are in the second part of the Seventh Circle, transformed into gnarled trees whose limbs are broken off by tormenting Harpies.

And in the Eighth Circle, the fortune tellers, prophets, and diviners, such as myself, who try to predict the future from earthly signs and wonders—and what is genomics but an earthly wonder?—must walk through eternity with their heads fixed on backward: "Because he aspired to see too far ahead / He looks behind and treads a backward path."

My colleagues and I decided not to try and convince the skeptical reviewer; instead, we sent the paper to another journal. The editor of that journal made suggestions about the title. Like our skeptical reviewer, she thought the title needed to be clearer about why all the behaviors we studied go together: "I think it would be good to include *something* about how you've identified a shared genetic factor among various traits." *Sin* is not a word you can use in *Nature Neuroscience*. We said we were studying the genetics of "traits related to self-regulation and addiction."

All this talk of sin—my Christianity is showing, the American, Southern, Protestant, Evangelical, "praise God and pass the ammunition" brand of Christianity into which I was born and raised. Or, rather, my ex-Christianity. I don't go to church anymore, don't pray anymore, don't identify as Christian anymore, don't believe in God anymore. At least, I don't think I believe in God anymore. I'm a scientist. But being raised Evangelical is like getting chicken pox as a child: You might recover, but you are never free. The virus will live in your nerves until you die.

Because I was raised Christian, I immediately recognized a

horrific theme in Travis's nightmarish trip. Because I was raised Christian, I know why studying the genetics of punished behaviors often unsettles people, like a half-forgotten nightmare with emotional tendrils reaching into the light of day. "I inherited a predisposition toward transgression" is a dark fable that's been told before. It is the doctrine of original sin.

Sin in Christianity is one word for two ideas, and the relationship between the two concepts of sin has been debated for most of church history. First, there are sin*s*. Lowercase "sins" are acts that contravene specific laws or standards. *Forgive me, Father, for I have sinned.* This type of sin is an action verb. This conception of sin is not specific to Christianity. People will disagree about which actions should be considered "right" versus "wrong." People will disagree about whether some actions are inherently wrong or, rather, are wrong because of the consequences they bring about. But regardless of their religious background, or lack thereof, most people believe that there are some things that people ought not to do, even if you are not accustomed to using the word *sin* to label the ought-nots.

Second, there is "original sin," or "hereditary sin," what the eighteenth-century U.S. American revivalist Jonathan Edwards called "the innate sinful depravity of the heart." I like contemporary British author Francis Spufford's definition better: Uppercase Sin is "the human propensity to fuck things up. . . . It's our active inclination to break stuff, 'stuff' here including moods, promises, relationships we care about, and our own well-being and other people's." The Christian doctrine of original sin is that this propensity to fuck things up is part of our intrinsic nature, what flesh is heir to, passed down from our ancestors Adam and Eve. They committed a sinful act by eating from the tree of the knowledge of good and evil. That act polluted their entire nature, and they bequeathed their fallen nature, their propensity to Sin, to all their descendants. Lowercase sin is an action; uppercase Sin is a condition.

One of the great theological debates in Christian history was between Pelagius, a monk who was eventually condemned as a heretic, and Augustine, a bishop who was eventually canonized as a saint. Pelagius argued that there was no such thing as hereditary sin. Sin was not a tragic inevitability, but something that was freely chosen. His view was summarized by the philosopher Paul Tillich in this way: "The contradiction of the moral demand must be an event of freedom and not a natural event." That is, you are legitimately punishable on account of what you have done, not what you are or what you have inherited. Conversely, if something is part of nature, then it cannot also be sin. In the words of Pelagius's supporter Julian, "What is natural cannot be called evil. . . . If you say it is a matter of will, it does not belong to nature; if it is a matter of nature, it has nothing to do with guilt." Pelagius and his followers endorsed an optimistic, if demanding, view of life: You have the freedom to choose. You will not be punished without the opportunity to make better choices.

Augustine, disagreeing with Pelagius and Julian, insisted on the primacy of inherited Sin, which, in his view, justified punishment in spite of its inescapability. In fact, Augustine insisted, belief in inherited Sin was necessary to maintain belief in God. The world, after all, is broken and full of pain. Even infants suffer. A good god would not allow the innocent to suffer, ergo, Augustine reasoned, even infants must not be innocent. And if infants aren't innocent, then guilt must begin in the womb. It must be something physically inherited. "The transmission of hereditary sinfulness is bound up with the reproductive process. . . . Infants arrive polluted by sin; since they have committed no actual sin, remission must be for the guilt attaching to a fault in their nature."

In Augustine's view, which became church orthodoxy, the only way out of deserving God's punishment is salvation. "Save" is how the English translators of the Christian New Testament translated the Greek word *sozo*, which can also mean to be healed from injury or sickness. Christian salvation is *bodily*. Jesus insisted that men must be reborn. But even salvation, to Augustine, was not something under your control; rather, it was dependent on God's freely given grace. Augustine's doctrine of original sin can be as terrifying as a bad acid trip: There is nothing you can do, or could have done, about being born with a sinful nature, but you are still blameworthy. Guilt is attached to a fault in your nature. You are to blame for yourself.

These Christian ideas about sin and salvation might seem irrelevant to you, my modern, enlightened reader, like taboos about curse words in a language you don't speak, spoken in a country you don't plan to visit. Or, if you were raised religious and fled the church as an adult, as I did, any talk of "sin" might feel like a traumatic reminder of the years you associate with being bullied and shamed. I understand: The Christian church would certainly consider me a sinner. I'm an apostate, after all, not to mention a divorcée and occasional drug user. But mercifully, I don't care *quite* so much anymore.

And yet.

Even if these Christian theological debates are couched in terms of God and Jesus, heaven and hell, faith and works, sin and salvation—that is to say, in religious terms that might seem inconsequential or off-putting to secular ears—I cannot help but find the questions at the heart of these debates compelling. What justifies blame and punishment? How free are we to avoid punishment? Are we to blame *because* we are free, or in spite of the fact that we are not? Is there a tragic element to human life, in which we are born with a predisposition toward those very acts that are punished? Is that tragic element unequally distributed at birth?

These questions are personal. Who among us has not felt, in some area of their life, like they were doing battle with their tendency to fuck things up? The apostle Paul: "I do not do the good I want to do, but the evil I do not want to do—THIS I keep on doing?" Who has not felt guilty and ashamed about things they also feel powerless to change? Who has not felt the undertow of their familial legacy—of trauma, violence, addiction, petty cruelty, narcissism, neglect, carelessness—and wondered if their predisposition to fuck things up was, in some part, not just modeled but *inherited*?

These questions are also political. What justifies blame and punishment? The Christian answer to that question is the doctrine of original sin, which says that fault is attached not just to our deeds but to our *natures*. The doctrine of original sin has shaped our family systems, our economic systems, our carceral systems. Remember Marcia Powell, cooked alive in the desert for the crime of drug use and sex work. That kind of torture needs a theology, what Nietzsche called a "metaphysics of the hangman."

And, finally, these questions are (re)surfaced by science. Science, like nightmares and acid trips, like horror movies and Greek poetry, takes us to uncanny places where the boundaries we urgently construct to make the world feel safe seem to blur and dissolve. Research like mine links inherited DNA differences between people with their tendency to behave in ways that invite punishment. Such science disturbs, provokes, unsettles. It needles us into remembering that we are not just minds; we are also bodies. We are moral agents embedded in an animal biology. We are not just the product of nurture but have natures, too.

My incarcerated correspondent is not the only one who has looked to behavioral genetics for potential absolution. Biology, and particularly genetics, is often used as an antidote for blame. Here is the old Pelagian wine in new genomic bottles: *The contradiction of the moral demand must be an event of freedom and not a natural event.* If something is a "natural event," then perhaps it is not a contradiction of the moral demand?

Americans, despite their puritanical roots, have a strong Pelagian streak. Often, the more "biological" they perceive a behavior to be, the less likely they are to perceive it as a moral failing. Take weight and appetite. People who believe that their weight is a product of their biology are less likely to blame themselves for having a larger body size, and scientific research on the biological causes of weight gain is frequently portrayed as antithetical to the idea that weight is a matter of moral responsibility. One article in *The New York Times* proclaimed that research on "still murky gene-environment interactions" disproved the idea that obesity is due to "laziness, gluttony, and sloth." Similarly, a *New Yorker* story about the weight loss drug Ozempic questioned whether the efficacy of the drug will "help us see that metabolism and appetite are biological facts, not moral choices." The framing of these pieces presumes a Pelagian either/or: Eating food can be understood either biologically, as a molecular process that can be altered by injecting yourself with a peptide that tricks your brain into thinking you have a full stomach, or morally, as a choice that can be blamed. By this logic, the "real" moral choices are the ones that can't be understood biologically, that are beyond the reach of molecular tinkering.

Or consider the television ads for antidepressant drugs. A little cartoon neuron releasing neurotransmitters talks to another little cartoon neuron, under the label "Chemical Imbalance." Viewers are urged to "Talk to your doctor about Zoloft!" People who suffer from mental illnesses like depression and who believe "biogenetic" explanations of their origins blame themselves less for experiencing mental health problems.

Autism is another example. In the middle of the twentieth century, autism was widely blamed on parents, particularly "refrigerator mothers," who were alleged to have damaged their children by being aloof, unaffectionate, and rejecting. (The refrigerator mother theory of autism was an outgrowth of psychoanalytic theories about women's sexual "frigidity," the symptoms of which were said to include wanting clitoral stimulation in bed and being insufficiently satisfied by one's maternal obligations.) In a 1960 interview, Leo Kanner, one of the early proponents of the refrigerator mother theory of autism, described the mothers of children with autism as "just happening to defrost enough to produce a child." The refrigerator mother theory of autism began to be discredited in the 1970s, coinciding with the publication of the first twin studies. One historical survey of research on autism described the turning of the tide: "Strong evidence against the unfounded view that autism results from neglectful parenting came in 1977 from . . . the first systematic, detailed study of twin pairs containing at least one child with autism."

More recently, my colleague Danielle Dick, also a behavior geneticist, wrote a parenting book encouraging parents to avoid blaming themselves *and* to avoid blaming their children by pointing at the role genetics plays in child development. "Just because your child is throwing huge fits doesn't mean you are doing anything wrong," she writes. "It just means your child inherited a temperament toward big feelings, feelings they are still learning to manage."

Body weight, depression, autism, tantrums—in these examples, the idea that behavior is genetically influenced is used to argue that someone is not guilty of being a wrongdoer. Other times, genetics is used to argue that someone is not the *victim* of wrongdoing. Eugenic thinkers in the United States and United Kingdom, for instance, have argued—wrongly—that people living in poverty are not being deliberately harmed by an unjust power structure but rather that poverty is an inevitable consequence of inferior genes.

Just as an emphasis on genetics can shift people's perceptions of whether a behavior is a moral issue, those invested in punishing a behavior as immoral often downplay the importance of biology. Consider sexual orientation: A 1992 article in *Christianity Today* by Joe Dallas, who to this day remains an outspoken opponent of gay rights, responded to one of the first twin studies of same-sex sexual behavior by criticizing the methods of twin studies and challenging the idea that genetics were informative regarding sexual morality: "When [scientific] inquiry is used to challenge traditional biblical teaching, then a clear balanced response from the church is called for. . . . Genetic origins do not justify sinful behavior."

Dallas was correct in one respect: Scientific inquiry into the biology of sexual orientation *did* undermine homophobia. An analysis of Americans' shifting attitudes about gay rights concluded that early-1990s media coverage of research on genetic influences on sexual orientation changed public opinion "in substantial ways." From the mid-1980s through the 1990s, highly educated liberals, in particular, showed rapid uptake of the belief that sexual orientation was predominantly due to biological causes, rather than to the learning environment or to personal choice. Conservatives, on the other hand, showed no such shift in their beliefs. In turn, people who believed that sexual orientation had biological causes showed greater increases in their positive feelings toward gays and lesbians and more support for gay rights over the course of the 1990s.

Drug use is another example. One highly publicized legal case in the United States involved a woman named Julie Eldred, who became addicted to opiate drugs as a teenager, stole jewelry to pay for fentanyl as an adult, and relapsed while on probation. Whether the court should be able to punish someone in active addiction for relapsing was the central question of the case. Eldred's possible genetic vulnerability to drug addiction was part of the background story emphasized by her defense lawyers: She had been adopted after being born to a mother addicted to cocaine.

One *amici curiae* brief submitted to the Massachusetts Supreme Judicial Court about Eldred's case sought to trivialize the contributions of behavioral genetics to our understanding of choice and blame. Comparing addiction to fatness, which they described as another "undesirable" and "self-destructive" outcome that is freely chosen, the authors wrote that "there is no necessary connection between heritability and involuntariness." Drug use should be punished as a voluntary choice, they argued. The court agreed.

All these examples reflect Pelagian reasoning: Biology is the antidote to blame; blame depends on minimizing biology. But how far can this reasoning be pushed?

In 2006, Bradley Waldroup got into an argument with his estranged wife, Penny, and told their four kids to "come tell your mama goodbye"—because he was planning to kill her. He ended up shooting his wife's best friend eight times and then slicing her head open. He then used a machete to slash Penny over and over. Remarkably, she survived. The assistant district attorney was appalled by the crime scene photos of Waldroup's trailer home in east Tennessee: "There are murders and then there [is] . . . hacking to death, trails of blood. I have not seen one like this. And I have done a lot."

At trial, a psychiatrist for the defense testified that Waldroup had a particular genetic variant on his X chromosome, as well as a history of child abuse, and that his genes and environmental risk combined to create a "dangerous cocktail" that predisposed him to violence. "A person doesn't choose to have this particular gene or this particular genetic makeup," he said. "A person doesn't choose to be abused as a child. So I think that should be taken into consideration when we're talking about criminal responsibility."

This argument was based on a study, "Role of Genotype in the Cycle of Violence in Maltreated Children," which was published in *Science* in 2002. The psychologist Avshalom Caspi and his colleagues genetically tested around 450 white men from New Zealand. Zooming in on the men in the study who had been abused or neglected as children (around thirty-five men), researchers found statistical evidence for a gene-environment interaction: The men who were maltreated as children *and* who had a particular version of a gene on the X chromosome were the most likely to develop violent behavior.

The study made headlines around the world. But over the next two decades, the scientific method that Caspi and colleagues used in their study became highly contested, both in the pages of academic journals and in the courtroom. The major problem, which was clear only in retrospect, was simple: The study did not include enough people for the statistics to be trustworthy. Today, genetic studies commonly enroll a few million people, not just a few hundred. If the specific gene that Waldroup had influences risk for violence at all, its effect in isolation is likely to be extremely small, perhaps increasing someone's likelihood of behaving violently by less than a percentage point.

(Individual genes, it should be noted, typically also have very small effects on autism, depression, appetite, tantrums, addiction, and sexual behavior—that is, on all behaviors for which the knowledge of genetic influence has been accepted as absolving. It is only when aggregating the effect of many genes that larger differences in risk emerge. The complex genetic architecture of human behavior is not specific to *violent* behavior.)

The jury in the Waldroup case could not know how much scientific controversy would unfold about the Caspi et al. study. They only heard that men who inherited a certain gene, particularly if they had also been abused as children, were more likely to be impulsively violent. Waldroup was found guilty and was convicted of voluntary manslaughter and second-degree attempted murder. He was sentenced to thirty-two years in prison. Prosecutors were "flabbergasted" that he was not convicted of first-degree murder, which could be punished with the death sentence. The genetic evidence appeared to be critical to the jury's thinking. As one juror described it, "Some people without this [gene] would react totally different than he would. . . . A bad gene is a bad gene."

The Waldroup case seemed to herald a new era of behavioral genetic evidence in criminal trials. Instead, the case turned out to be the exception, not the rule. Since 2007, information about a defendant's genetics has been entered into evidence in several dozen criminal cases, most commonly as a potential mitigation factor in sentencing for capital crimes, and this evidence has rarely worked out in the defendant's favor. In 2021, the New Mexico Supreme Court concluded that the type of genetic evidence introduced in the Waldroup case—"evidence of mere genetic susceptibility" on the basis of a single common gene—"is not relevant on the issue of deliberate intent." Despite these failures, defense attorneys are unlikely to lose interest in genetics completely: "In high-stakes criminal trials, defense attorneys will likely pursue any approach that could mitigate culpability, and thus they are likely to continue to introduce complex genetic evidence that could have such an effect. This is especially likely to be true in capital sentencing."

For autism, addiction, appetite, depression, and sex, science connecting genes to human behavior has tended to soften people's blame and condemnation. But violence is a different beast. Introducing genetics when someone has committed a violent crime is not reliably mitigating. In fact, some studies have found that people are *more* punitive toward violent offenders when they believe that a predisposition to violence can be transmitted physically from parent to child. Sometimes, judges, juries, and ordinary people are presented with arguments that defendants have a "fault in their nature"—and like Augustine, they think that the defendants should be punished anyway.

But *should* they be punished anyway? If we blame people for their behavior and their genetic inheritance helps shape behavior, then what does genetics have to do with blame? This is an ancient and enduring question. This is the question the man who wrote me from prison is asking me to consider. This is the question I am circling.

In the case of my correspondent, I can say with certainty that he did it: He kidnapped a woman and sexually assaulted her. I can also say with certainty that I find the conditions of the U.S. prison system abhorrent. No human—no animal—deserves to be caged in solitude, to live with the ever-present risk of violent victimization, to be denied basic care.

I can say with certainty that some people are born with certain genes that make it more likely for them to do things that *are* punished. (Whether or not they *ought* to be punished is a different question.) Children who lie and hit and defy are spanked or grounded by their parents. Teenagers who start fights and break rules are suspended from school. Adults who use drugs are put in jail. Men who commit violent crimes are imprisoned. I personally have published scientific papers on the genetic causes of all of these behaviors—childhood conduct problems, adolescent delinquency, adult substance use disorder, adult violence—and I am just one scientist.

I can say with certainty that I've committed plenty of sins. I am not writing letters from jail, but not because I'm innocent of any crime. I could, for instance, drive through Texas with LSD in my possession without any real fear of being stopped, searched, or arrested by the police, because of the unfair privilege afforded by my race and education. I can say with certainty that I find the disproportionate punishment of people who are poor, who are mentally ill, who are racial minorities, unjust.

What I am uncertain about: if anything about one's genetics matters for whether and how they should be punished for the sins they commit. Or if we should even see them as "sins" in the first place. Does science let us off the hook—or, at least, some of us off the hook some of the time? Or is science, in the end, as accusatory as a bad acid trip, as a fire-and-brimstone preacher?

In her memoir, *The Uninnocent: Notes on Violence and Mercy*, the lawyer Katharine Blake writes of her own uncertainty after her teenage cousin, Scott, in the throes of what appeared to be a psychotic break, brutally murdered a younger boy with a box cutter. The victim had been out riding his bicycle with his family; his mother was just far enough behind on the trail to be powerless to intervene. Scott was sentenced to life in prison without the possibility of parole. Considering his blameworthiness, Blake writes: "I felt certainty diminish. It was one thing to shout *Freedom for the prisoners,* another to proclaim it in the particular. Did Scott deserve freedom? ... At first this growing uncertainty made me quiet. But then I figured out that uncertainty wasn't only an absence, empty air and space, but a thing with its own shape and texture." She concludes that her story is one about "losing, finding, losing again, and searching for certainty." Instead of parsing the metaphysics of the hangman, she is seeking a "theology of possibility."

In contrast to Blake's searching humility, some of my fellow scientists are all too eager to proclaim their certainty. Some are absolutely convinced that scientific knowledge about the causes of human behavior, including its genetic causes, renders moral responsibility and punishment obsolete. My prison correspondent does not deserve to be punished, nor does Bradley Waldroup, nor does Blake's cousin Scott. Here, for instance, is Robert Sapolsky in *Determined: A Science of Life Without Free Will*: "We need to accept the absurdity of hating any person for anything they've done; ultimately, that hatred is sadder than hating the sky for storming, hating the earth when it quakes, hating a virus because it's good at getting into lung cells. This is where the science has brought us as well."

Others are absolutely convinced of the opposite conclusion: Science has nothing to tell us about how to treat one another, or it just confirms the rightness of our current practices. Kevin Mitchell, for example, after contemplating some of the world's most cutting-edge research in genetics, neuroscience, and physics in his book *Free Agents: How Evolution Gave Us Free Will*, concluded that he saw "no reason . . . not to continue to hold people responsible for their behavior in the ways we always have"—as if there is only one "we," as if we have "always" held one another responsible in the same ways.

As a recovering Christian, I am suspicious of men who confidently declare they have definitive answers to complicated questions about what it means to have a body. For me, having an animal biology sometimes feels like a curse and sometimes feels like a relief. I love things that are bad for me. I punish myself, and other people, for things we don't know how to change. I no longer believe in Jesus, but I do believe we inherit predispositions toward sin. Sometimes I am outraged and convinced that someone deserves punishment, only to find my convictions shimmering, fading, as I inspect them more closely. I have spent much of my adult life studying how our genes cause us to behave in ways that are praised or punished, and I am still trying to make sense of what this work means for how we should treat one another.

In this book, I think through that question. As you will see, I offer few definitive answers. I will not neatly resolve enduring mysteries about the human condition in a single book. My ambition is more modest, which is to ponder these mysteries from the somewhat unusual vantage point of someone who is intimately familiar both with the scientific study of child development and behavioral genetics and with the Christian doctrines of sin that have profoundly shaped our legal and cultural traditions regarding blame, punishment, and human nature.

The ex–Evangelical Christian community uses the word *deconstructing* to describe the long and often painful process of picking apart the stories you were taught as a young person growing up in the church. What were you taught to believe, and how does that belief still implicitly shape your behavior, relationships, and identity, even if you haven't explicitly thought about the belief, and the stories behind it, for years or even decades? As an ex-Christian, I've been deconstructing one set of stories: religious myths about what we inherit and what we deserve. And as a behavioral geneticist, I've been deconstructing another set of stories: newspaper headlines and parenting books and political messages and therapeutic pronouncements and criminal trials, many of which position biology and morality as existing on opposing sides of a binary divide. Like the Christian stories, the scientific ones are also trying to answer some big questions: Why does a person behave as they do? Who is to blame, and for what? Who deserves to suffer, and why? And on closer examination, many of the supposedly scientific stories turn out to be modern retellings of, or reactions to, the ancient Christian myths. What I once thought were two deconstruction projects are actually one.

Deconstruction can be horrifying. Deconstruction tears down the boundaries that define our neat binaries—biological versus moral, animal versus human, unwitting versus chosen, innocent versus guilty—and binaries are comforting in their simplicity. But binaries are also confining. We chafe against sharp, artificial edges. We drop acid and watch horror movies and read Greek plays and conduct scientific research on purpose, because we hope that drugs and art and science will be powerful enough to dissolve the boundaries we ordinarily construct to feel safe. We want to be changed. I think we need to be changed. We need a suppler understanding of what it means to be responsible for ourselves, and responsible to each other, than can be found in the dichotomous eschatology of heaven versus hell. My hope is that by deconstructing some of the binaries that encage our stories about human behavior, we can grow more elastic, more creative in our thinking about how each of us deserves to be treated.

While I was deep in the writing of this book, I took a walk with my son, who was then still in elementary school. Apropos of nothing, he said, "If you thought reality was only two dimensions and started walking east in a straight line, you'd be confused when you eventually ended up west of where you started." I don't know where he might have learned this idea, or what prompted him to tell me about it. ("Suffer the little children," Jesus said, "for such is the kingdom of heaven.") But his comment struck me as a truer, wiser encapsulation of the relationship between biology and moral responsibility than all the confident declarations that east can never be west, that science obliterates blame or science is irrelevant to blame. What if humans and animals, morality and nature, sin and inheritance, accountability and mercy do not occupy opposite sides of the spectrum, because the spectrum bends back on itself?

In this book, I invite you to circumnavigate with me. I want to show you how if you travel east far enough you end up west, how every sinner was once a child, how every human who feels herself to be a causal agent is a consequence of forces, including genes, beyond her control. How bad genes can be good; how nature and nurture collapse into each other; how punishment is just as much part of our evolutionary inheritance as the sins that are punished. If you follow the science far enough, you might worry you're falling off the map, but only because the map flattens reality. We might progress through contraries.

Letter

An interlude, in which I pose to the reader the questions I was asked by a man imprisoned for violence

When they say "the apple doesn't fall far from the tree" how do you interpret that statement?

Why would a young boy of sixteen attack a total stranger, a female, at knife point in broad daylight at a busy intersection and make her drive against her will to sexually assault her? What would drive a boy to do such a thing?

What makes a child go bad? Is it nature or nurture?

Luck

In which I conceive a son, and fear that the unlucky seeds of a violent constitution have already been planted in him

LET ME TELL YOU a fable.

Once upon a time, a man and a woman lived in a beautiful garden. Their God told them that they could eat of any tree in the garden, except for one. "You must not eat from the tree of the knowledge of good and evil, for when you eat from it you will certainly die." But one day the woman noticed that the fruit of the tree looked good to eat and desirable for gaining wisdom. So she picked some fruit, and ate it, and gave some to the man.

Then their eyes were opened, and they realized they were naked. They sewed fig leaves to cover their bodies. Their God was angry, because they had become too much like him, knowing good and evil.

The God told the woman that he would make childbirth painful for her, and that she would feel desire for her husband, but he would rule over her. And the God banished the woman, and her husband, from their garden, their home.

The man named the woman Eve, and she became pregnant with a son.

The beginning of my son's birth is deceptively easy. Labor starts with the onset of what feels like bad menstrual cramps: a wringing in my lower belly, a dull ache through the sides of my hips and lower back. For the first few hours, E, my first husband, sleeps unaware in our bed. Out in the living room, I time my contractions with an iPhone app and read Martin Amis's *London Fields,* a novel about a murder that has been foretold by the victim. Like me, the characters brood and panic and joke about the impending "horror day" that can be anticipated but not prevented. The novel's narrator observes: "You can't stop people, once they *start.* You can't stop people, once they *start creating.*" I've started creating a child, and now there is no stopping.

I wonder if I am in active labor—which obviously I am not, because there is still an "I" with enough presence of mind to time my contractions and read a novel and wonder if I am in active labor. In early labor, there is still the familiar executive self, the ego, who guides my actions and reflects on their consequences. "Here I am, double-checking my hospital bag. Here I am, wondering if that twinge is strong enough to count as the beginning of another contraction." I was naïve enough to imagine, briefly, that labor might carry on like this. Painful, yes, but recognizable.

Early labor is bad cramps. Unmedicated active labor is a hot-air balloon being inflated inside your pelvis. You will have to stretch to contain the whole balloon, the burners keep firing, the balloon keeps rising, you will break open, the expansion will break you—and still you expand.

I took the requisite hospital birthing classes, in which the nurse showed us timetables and size comparison charts. Seven centimeters is about the size of the top of a soda can. If your cervix is seven centimeters dilated, you are transitioning into the last stage of labor. Ten centimeters is about the size of a bagel. If your cervix is ten centimeters dilated, then it's just large enough for a child's head to be extruded. I am good at remembering facts and figures, but academic preparation is, in this case, no preparation at all.

The philosopher Thomas Nagel once asked, in an essay about consciousness, "What is it like to be a bat?" His argument was about the subjectivity of conscious experience: There is something that it is like to *be* a bat, and objective knowledge of the bat's nervous system will be insufficient for understanding the bat's subjective internal reality. Knowing how echolocation works does not translate into an understanding of what it is like for the bat to echolocate. Attending hospital birthing classes, it turns out, is like visiting the bat cage at the zoo in preparation for turning into a bat. Knowing how the uterus contracts and the cervix expands does not translate into an understanding of what it is like for the woman to labor.

Nagel questioned whether there could be any way to imagine the inner life of the bat, who is fundamentally alien. I don't know what it's like to fly, but I know what it's like to try to push your young out of your vagina. Childbirth expands you and expands your range of imagination about what it is like to be an animal.

No, childbirth is when you give up imagining that you aren't an animal.

The labor and delivery room is full of strangers. I would ordinarily feel ashamed to be half-naked and grunting in front of people I don't know. But shame is a self-conscious emotion, and self-consciousness requires a splintering of experience: To feel shame, one part of the self must hold itself at a remove, observing the other parts. Judging the other parts. To realize that you are naked, your "eyes" must be open. That observing/judging part of me ordinarily insists that she *is* I, my true self. In labor, however, that imperious part of me, which ordinarily seems so essentially *me,* is gone. This is just as well, because it turns out that "I" am not necessary for childbirth. My body, which "I," frustrated by its needs and limitations, so often treat with scorn, doesn't need "me." The body has work to do.

Is this lack of self-awareness what it was like for Eve in the garden? Before her crime, Eve lived a life without shame. She did not realize she was naked. After she ate the forbidden fruit, her eyes were opened. She brought consciousness to humanity. A few experiences—sex, childbirth—can still momentarily dissolve her shame, restore the animal unity of her experience, shut the eyes of her observing/judging self, make her forget she is naked. God's curse is that these very same experiences will also make her suffer.

After more than twenty-four hours of labor, and after I've been pushing for nearly four hours, my son won't come out. His head is wedged in my pelvis; his heart rate is dropping. They inject me with adrenaline to jump-start his heart. My doctor tells me that she has no choice but to operate, and the nurses start prepping me for an emergency cesarean. But something is wrong with the epidural they have put in; they cannot get me numb enough for surgery. The observing self returns, sees the pain to come, sees that my son might die, or I might, or we both might. I am terrified. I sob, "Please don't cut me! I can feel everything." I also sob, "Do what you have to do. Cut him out. Please don't let him die!" There is not enough time to take the needle out of my back and replace it. Instead, they slap an anesthetic mask over my face. Now all of me is gone, not just the ego.

I miss the first few hours of my son's life. I am underwater, kicking hard toward the surface of consciousness, but unable to escape the undertow of anesthesia. When I finally wake, E is holding our son against his chest. Like all newborns, he looks like an ancient, wizened creature. The nurses say, "He's perfect." I examine him as I clumsily, painfully try to breastfeed. He *is* perfect, to me. He is *alive*.

He also has, I soon notice, bilateral partial syndactyly, i.e., webbing between his second and third toes on both feet.

The woman, Eve, gave birth to her first son, Cain. Then she gave birth to his brother, Abel.

Cain and Abel both brought sacrifices to the God who had exiled their parents from their garden home. But the God did not look on Cain's offering with favor. He instead favored his brother, Abel. Cain was angry and depressed. He invited his brother out into the field, attacked him, and killed him.

The God cursed Cain and sent him away to be a restless wanderer throughout the earth. Cain despaired that he would no longer be in the presence of his God. Having killed, he was certain he would now be killed. But the God put a mark on Cain, so that no one who found him would kill him.

Toe webbing has been called a "mark of Cain," because of its correlation with aggressive behavior. The technical term is "minor physical anomaly," or MPA. MPAs are seemingly inconsequential. You might not notice them if you aren't looking closely: webbed toes, hair whorls (where hair grows in a spiral pattern), low-seated or asymmetric ears, a tongue with deep grooves, a curved fifth finger. But MPAs are a subtle sign that something might have gone awry during fetal development, perhaps because of a genetic mutation or perhaps because of some environmental insult, like maternal infection or exposure to alcohol. The brain and spinal cord develop from the same layer of the early embryo as the skin, nostrils, hair, and nails. The same errors in fetal development that leave their visible mark on the toes, palate, or ears might then also produce unseen anomalies in the developing nervous system.

Researchers first noticed decades ago that boys who had more MPAs at birth were more aggressive and hyperactive, on average, when they were preschoolers. As children grow, the manifestations of aggression change: Aggression at age three means biting another child at daycare. At nine, it means hitting and scratching another child on the playground. At seventeen, it means assaulting another person or threatening them with a weapon. The form of aggression changes, but the relationship between aggression and MPAs stays the same. Around the world and across ages—among low-income French Canadian boys, among Danish young adults, among African American teenagers—young men with more MPAs are more likely to commit violence. The MPA-violence relationship is particularly strong among young men who also face environmental risks, such as an impoverished home environment. And at the extreme end of the violence spectrum, case studies of serial killers have noted that some have unusually high numbers of minor physical anomalies. Carlton Gary, a.k.a. the "Stocking Strangler," who was convicted and executed for raping and strangling multiple elderly women, is reported to have had at least five MPAs, including fully attached earlobes and webbed fingers and toes.

But why was I scanning my newborn child for MPAs? All pregnant women fear something; throughout my pregnancy, I feared that, like Eve, I would give birth to a Cain. Before conceiving, I interpreted Eve's curse—*with pain you will give birth to children*—as referring only to the physical pain of childbirth. But some translations of the Bible render that word not as "pain" but as "sorrow," and now I know: The sorrow, pain, and fear of motherhood begin long before the first contraction and continue long past the last one. In the third chapter of Genesis, God curses Eve with sorrow in childbearing; in the fourth chapter, Eve's child kills her other child. Did Eve know, as her God was shaming her, that this was part of her curse, that her capacity for creation would also hazard destruction?

Despite this being a major plot point in the first book of the Bible, none of the mommy blogs and parenting books I browse include sections on parental concerns about giving birth to a son who will grow up to be violent. The modern American parenting-industrial complex relentlessly preaches the good news that motherhood can be made safe, if only you mean well enough, strive hard enough. The books and podcasts and social media influencers, the wooden toy kits and bento box cookbooks and alternative vaccination schedules—all of it is oppressively optimistic about the ability of mothers to perfect their pregnancies, their selves, their children. If we *Expect Better*, if we *Respect Babies*, can we guarantee that our children will be not just physically healthy and cognitively bright but also morally good? God himself could not ensure that his children would not fall into sin, but the bookstore offers you the opportunity to *Raise Good Humans*, *Raise Lions*, *Raise a Feminist Son*, or, at the very least, *Raise Kids Who Aren't Assholes*. I read these books, too, and learn from them. But they cannot lull me into a feeling of safety. By conceiving a child, I feel like I have gone all in at the roulette table. I am exhilarated, remorseful, eager, terrified at how extravagantly I have exposed my life to brute luck.

Instead of reading parenting books, I found myself, without quite realizing what I was doing, seeking out versions of the Genesis fable. I wanted to read about flawed women who gave birth to violence. In my eighth month of pregnancy, I went to the movie theater to watch Ridley Scott's *Prometheus*. In the opening scene, a humanoid creature drinks a black liquid, which causes it to disintegrate, sowing the earth with its seed. Later, Elizabeth, a scientist, has sex with her partner, Charlie, who has been secretly dosed with the same black liquid, which has been recovered from an alien planet. She becomes pregnant and undergoes a self-administered cesarean in a high-tech medical pod. She births a writhing monster. One of the most horrifying moments of the movie is the least fantastical: To be delivered from her pregnancy, she is cut open and stapled back together. Despite having been bisected so soon before, she must walk, run, stumble, get on with things. She must cope with the destruction wrought by the creature she has birthed. As I was waddling out of the theater, a young theater attendant stopped me: "Please tell me you didn't just watch that."

Of course I did. I was seeking catharsis. Aristotle introduced the word *catharsis* in his *Poetics* to describe the audience's response to tragic plays. In Greek tragedy, the hero makes a fatal error, which is not initially obvious to either the hero or the audience but is later revealed, and he experiences a reversal of fortune. In *Prometheus,* for instance, Elizabeth makes an error by having sex (which in a horror movie usually portends doom) and conceives a monster. Tragedy, according to Aristotle, provokes feelings of pity and fear—pity for the hero, who meets a horrible end, and pity and fear for ourselves. We, too, will unwittingly make errors, which might prove to be our downfall. We, too, are subject to dramatic reversals of fortune.

Aristotle developed the idea of catharsis to describe what made effective tragedy; Freud later adopted the term to describe what made effective therapy. How could, he wondered, the "talking cure" work to alleviate people's hysterical symptoms—their mysterious compulsions, paralyses, fainting spells? In Freud's view, psychoanalysis, like Greek tragedy and modern horror movies, was a contained environment where people could bring emotionally intense, even traumatic material into conscious awareness so that they could learn to live with it. Unlike the theater attendant, a psychoanalyst would not be surprised that I was watching a gruesome onscreen cesarean mere weeks before my due date. I was there to allow myself to feel afraid.

Elizabeth's fate is fantastical, but there are more realistic versions of the same story. While folding baby clothes, I watched *We Need to Talk About Kevin*, based on Lionel Shriver's novel of the same name. The main character is Eva, a name I doubt is coincidental. She was once a successful writer, but now she is a mother, and she cannot bond with her baby. In one scene, she pushes his baby carriage next to a construction crew, visibly relaxing at the sound of the jackhammers, a reprieve from her son's incessant crying. "Mommy was very happy before little Kevin came along. Now every morning mommy wakes up wishing she were in France!" As Kevin grows up, he is increasingly aloof, narcissistic, and cruel. Right before his sixteenth birthday, he murders his sister, father, and classmates with his bow and arrows. Watching the final scenes, when Eva visits Kevin in prison, I felt my son kick in my belly.

That same month, I read *Far from the Tree: Parents, Children, and the Search for Identity*, by Andrew Solomon. Solomon interviews Sue and Tom Klebold, the parents of Dylan Klebold, who was a gunman in the 1999 school shooting at Columbine High School in Colorado. Klebold and his friend Eric Harris killed thirteen people and set up propane bombs intending to kill as many as six hundred. Solomon presents the Klebolds as normal, sunny, and suburban. Sue mentions that she has watched the Roman Polanski movie *Rosemary's Baby* and sympathized with Rosemary, who discovers by the movie's end that she has given birth to Satan's child.

Sue agonizes: "I can never decide whether it's worse to think your child was hardwired to be like this and that you couldn't have done anything, or to think he was a good person and something set this off in him." There is no ironclad evidence in Dylan's case history to suggest that he was "hardwired" to be a school shooter. There is only the gaping hole in our ability to explain his violent behavior with any aspect of his warm and advantaged home environment. The Klebolds are left wondering: Could things have gone differently? If so, how? Is it worse if they could have averted their child's fate, or if they couldn't have?

The question the Klebolds ask themselves is one I pose to you: If your child committed a heinous crime, would it be worse for you if you learned he was "hardwired" to be violent regardless of his environmental circumstances, or if you learned his behavior could have been prevented by something you didn't do?

Sue's agonizing question is also posed in *Hereditary*, a movie I avoided when it came out but later watched during my third, and final, pregnancy. In an early scene, the adolescent Peter, who will soon discover how his grandmother's sins have condemned his family to possession, insanity, and death, discusses Sophocles's tragic play *The Women of Trachis*. The teacher asks the class, "Sophocles wrote the oracle so that it was unconditional, meaning Heracles never had any choice, right? So does this make it more tragic or less tragic than if he had a choice?" Peter has no answer.

All these movies, books, and plays are what the philosopher Bernard Williams, in an essay about *The Women of Trachis*, called "stark fiction." The main characters are well-meaning strivers, but they are not spared from suffering. Stark fiction, in

Williams's view, functions like a dream, or like psychoanalysis: It allows us to tolerate truths that are ordinarily too painful to contemplate for very long, if at all. In Williams's words, the stories I seek out when pregnant reflect "the very plain fact that everything that an agent most cares about typically comes from, and can be ruined by, uncontrollable necessity and chance." It would be impossible to live our lives if we were always thinking about how we could be ruined by chance, but pregnancy confronts me with the hideous contingency of every aspect of existence—indeed, of existence itself.

Of course, in my search for Eve's story, I need not content myself with novels and movies. Science has stories as horrifying as fiction.

In one Dutch family, first described by the scientist Han Brunner in the early 1990s, all the women were leading ostensibly normal lives: They had gotten through school just fine; they had relationships and jobs and apartments. They were free of major psychopathology, except for the fact that some of them had been traumatized by their interactions with their brothers, sons, uncles, cousins. But the men in the family were not okay. Investigators evaluated five men in the family and obtained detailed histories on three more. They were unremarkable in terms of their gross physiology; their height and weight were normal. A phrenologist examining their skulls would have found nothing of note. But these men also suffered from frequent night terrors and borderline intellectual disability. And they were all repeatedly and seriously violent.

One member of the family raped his sister when he was twenty-three and was then incarcerated in a unit for psychopaths. When he was thirty-five, he stabbed one of the prison wardens with a pitchfork. Another member of the family was chastised by his boss for not doing well at work. He tried to run his boss down with a car. A third member of the family forced his sisters to undress at night, threatening to cut them with a knife. Two other members of the family were "known arsonists."

A team of geneticists led by Brunner discovered that "affected" males in this family all had a mutation in the *MAOA* gene, which is on the X chromosome. Women have two X chromosomes, which can buffer them from the consequences of a mutated X-linked gene. Men, however, have only one X, so if their version is mutated, the effects can be catastrophic. Unlike most "common" genetic mutations, which are found in more than 5 percent of people and typically have very small effects on behavior, rare mutations like the one found by Brunner and his team can have extremely large effects. A woman who carries one copy of this mutated X-linked gene has a fifty-fifty chance of passing that version on to her male child.

Heads: Your son leads a "normal" life. Tails: Your son is intellectually disabled and incorrigibly aggressive. How much would you wager on a coin flip?

The different effects that sex-linked mutations can have in women versus their sons reminds us: Children are not our copies. In *Far from the Tree*, Andrew Solomon put it perfectly: "There is no such thing as reproduction. When two people decide to have a baby, they engage in an act of production, and the widespread use of the word *reproduction* for this activity . . . is at best a euphemism to comfort prospective parents before they get in over their heads." An occupational hazard of being a behavioral geneticist is that I am not comforted by euphemisms. I know too well, as a prospective parent, that I am producing something new and unknown. I am wagering much on the outcome of genetic coin flips.

When I was still a fetus in my mother's womb, all the genes I inherited from her and my father were shuffled up to make tiny egg cells, each one with its own unique combination. And my eggs are trifling in number in comparison to the billions of sperm produced by a man. As a result of this combinatorial explosion, there are trillions upon trillions of possible genetic combinations my husband and I could have produced when we conceived our son. If every single one of our potential genetic children was a dollar bill, the stacks of dollar bills would reach to the moon and back a dozen times. Any prospective parent should consider: What potential lives are in that stack?

What a child is inheriting, when they are conceived, is an entirely unique combination of 3 billion DNA "letters," which are commonly abbreviated A, C, G, and T. A three-letter sequence of DNA letters, what you might think of as a genetic word, is called a *codon*. When "coding" DNA is "read" by the cell, a molecule called messenger RNA carries copies of the DNA sequence to specialized organelles, called ribosomes, which turn the DNA instructions into proteins. A three-letter DNA codon corresponds to a particular amino acid, which is the building block of a protein.

The *MAOA* gene, for instance, codes for an enzyme called monoamine oxidase type-A, which "eats" molecules that the cell doesn't need anymore. Ordinarily, the MAO-A enzyme breaks down the chemical messengers that neurons use to communicate with one another, messengers like serotonin or norepinephrine. Most psychiatric drugs, like Prozac or Ritalin, work by changing levels of these chemical messengers. But in the Brunner syndrome family, a single DNA letter in the *MAOA* gene had been changed to a T instead of a C. This changed the codon from CAG, coding for the amino acid glutamine, to the codon TAG, which is the code for *STOP*.

Stop adding new amino acids; stop making the protein. Stop regulating one's rage; stop developing the joint excellence of reason and character.

The resulting MAO-A enzyme is a stunted, deformed thing that cannot function normally. MAO-A enzyme deficiency, in turn, leads to a buildup in the neurotransmitters that neurons use to communicate with each other, a suppression of rapid eye movement (REM) sleep, and a tendency toward explosive outbursts of anger, especially in response to fear or frustration. To inherit this mutated gene was to be prone to engage in arson, rape, assault, or even homicide. There is, at present, no cure.

In the decades since Brunner and colleagues' original articles about the Dutch kindred, additional mutations in the *MAOA* gene have been detected in just a handful of other families, whose identities remain protected. We get but glimpses of their lives in published case reports, but even those fragments attest to sorrow. In Portugal, a man is arrested in his midtwenties for attempted murder. While he is in prison, his younger brother is diagnosed with Brunner syndrome, so he is genetically tested, too. We do not hear from the case report what happens to his prison sentence; we know only that at the age of thirty-one, he is committed to a psychiatric hospital for homicidal fantasies.

In France, a seven-year-old boy, with young, healthy parents and a typically developing younger sister, was referred to a genetics clinic after he was diagnosed with an autism spectrum disorder. He was slow to walk, slow to talk, slow to learn to read; he was easily angered when frustrated and liked to play with fire. The geneticists discovered that he had inherited a previously unknown mutation in his *MAOA* gene, a slightly different error than found in the original family, but one with life-changing consequences. In Italy, another boy, five years old, was evaluated because of speech and language delays, possible symptoms of autism, hyperactivity, seizures, poor tolerance to frustration, and aggressive outbursts. His genetic testing results revealed yet another new mutation in the *MAOA* gene.

Two brothers, both in their fifties, were referred to a genetics clinic in Australia, where they appeared passive and compliant during their clinical interviews. Only later did the investigators hear from their families about their history of school expulsions, property damage, physical aggression toward family members. Another pair of brothers, both in their thirties, were referred to the same clinic; they also have a history of impulsivity, aggressive behavior, and diagnoses of ADHD. The two pairs of brothers turn out to have different *MAOA* mutations. The scientists publishing the case report are surprised to find two families in one area who are suffering from the same rare disease. They wonder: What if Brunner syndrome is not as rare as people imagine? How many other angry, impulsive, socially and intellectually impaired young men with genetic abnormalities go undiagnosed? How often, after all, when affronted by someone's badness, are we motivated to search for the biological mechanism that might have caused it?

The case reports of families affected by Brunner syndrome are, like the plots of *Hereditary* and *We Need to Talk About Kevin* and *The Women of Trachis* and the Book of Genesis, horror stories. Horror stories allure us because they allow us to sidle up to ideas that we find too disturbing to confront directly. They allow us to engage with our fears but allow us to disengage just as readily. In the case of rare genetic diseases, like Brunner syndrome, we can reassure ourselves: That is a one-in-a-million event; that wouldn't happen in my family. So, too, can we remind ourselves, when the novel or movie is over: It's not real. It's only make-believe. When we wake from a nightmare or return from an acid trip, we give ourselves the same comforting reminders: It was only a dream, it was only a dream, it was only a dream.

But wait. Don't disengage yet. Don't reassure yourself yet. Sit with me.

Consider: What ordinary truth is revealed by these stories, which are ostensibly about the imaginary and the extreme?

Realize: There are thin places in human life, where the boundary between bad luck and bad behavior blurs and dissolves.

Philosophers have a technical term for what befalls Eva's child in *We Need to Talk About Kevin*, for what has befallen the men who inherited a defective *MAOA* gene and killed another person in a fit of frustrated rage, and for what I fear will befall my unborn child: moral luck. Someone is affected by moral luck when what they do depends, in some significant way, on factors beyond their control, but we continue to treat them not just as unlucky but as *bad*, as an object of moral judgment.

Imagine two drunk drivers, equally impaired, equally reckless. One strikes a pedestrian walking home late at night; the other would have been unable to avoid striking that pedestrian, but there happened to be no pedestrian to hit. Many would balk at punishing the second driver as severely as the first, but the only difference between them is something—the presence or absence of another person—that is beyond their control. The drunk driver who hits someone is treated not just as more unlucky but as *worse* than the drunk driver who doesn't.

Imagine two women, equally bored by life, equally restless. One gives her boyfriend LSD and he has a bad trip inside the safety of their hotel room. The other gives her boyfriend LSD and he, in the throes of paranoid delusion, runs out of the hotel room into the surrounding West Texas desert, where he rapidly suffers from heat exhaustion. Only one woman would be blamed for contributing to a man's death.

Consider: When in your life have you gotten lucky and so were spared from blame?

Realize: "You are poised on fortune's razor-edge."

The drunk driver and the drug-using girlfriend are examples of what philosophers call "circumstantial luck." They illustrate how commonplace it is, in our everyday life, to treat people as morally blameworthy for outcomes that were, to some significant extent, due to factors not under their control. Science and fiction can re-weird the commonplace so that we pay new attention to its darkness.

Philosophers sometimes argue with each other by writing miniature science fictions and then describing how they feel about their fictional characters: "My intuition is that Thoroughly Bad Chuck is indeed morally responsible for committing murder." They then argue about how their fictional character is, or is not, logically different from some other fictional character. "This is how Thoroughly Bad Chuck is different from Occasionally Bad Susan." The point of this exercise is to trace the hidden logic behind our moral intuitions and to perhaps change those intuitions by pointing out logical inconsistencies.

One such science fiction, by the philosopher Al Mele, imagines an evil superhuman goddess Diana, who creates a zygote, Z, within the womb of a woman named Mary: "She combines Z's atoms as she does because she wants a certain event E to occur thirty years later. From her knowledge of the state of the universe just prior to her creating Z and the laws of nature of her deterministic universe, she deduces that a zygote with precisely

Z's constitution located in Mary" will develop in a particular way. The zygote is born; the child Z develops into a man named Ernie, who does indeed act in a way that brings about the intended event.

Mele then asks: Is Ernie responsible for the event, despite being deliberately created to make that event come about?

These types of stories about "manipulated agents" focus our attention on what it means for human behavior to be caused. Ordinarily, when we are living our lives in the first person, we feel ourselves to be the causal agent who decides between courses of action based on reasons. What is it like to be a person doing things? It feels like you are making things happen. We then impute this subjective view of behaving to others: They, too, are deciding, causing, making things happen. Stories about manipulated agents are designed to trigger a shift in perspective, from the subjective to the objective. Now we are asked to see an actor and her actions as just one more component in the larger flux of events, as a phenomenon that is itself caused.

We can imagine an alternative version of the Ernie scenario: "An evil geneticist uses CRISPR-Cas9 technology to edit a zygote's genome to have the *MAOA* mutation discovered in Brunner syndrome. She does this to create someone who will be violent. Thirty years later, the zygote is now a man named Ernie who stabs someone with a pitchfork."

This scenario is science fiction, but it could conceivably become a real scientific possibility. In 2018, a Chinese scientist, He Jiankui, made international headlines for editing the genomes of twin girls whose father was HIV-positive. He targeted the *CCR5* gene, which codes for the protein that HIV uses to enter human cells, in an attempt to make them resistant to HIV infection. Although that goal sounds worthwhile, the procedure was risky for the children: It is still unclear whether the gene editing technique also made "off-target" changes to other parts of the girls' genomes, causing cancer or earlier mortality, which could then be passed down to their children. And he was working without any guidance or approval from the ethical agencies that govern the use of experimental medical treatments or genetic technologies. He Jiankui was condemned by the international community of scientists and bioethicists, fired from his job, fined, and imprisoned, but I doubt he will be the last rogue scientist who uses gene editing in ethically dubious ways.

Which brings us back to our updated zygote scenario: If a rogue gene-editing scientist, using her knowledge of the laws of nature, deliberately created a zygote so that it would develop into a violent person, would that person be morally responsible for what he does?

My first intuition is to say no, Ernie is off the hook. He is a dangerous but pitiable figure who needs to be contained to prevent future harms, not an evil and deplorable figure who deserves to suffer for his sins. *"Tout comprendre, c'est tout pardonner."* To understand all is to forgive all.

But on second thought, I'm not so sure. The shift in perspective, from subjective to objective, from "He stabbed someone with a pitchfork" to "A genetic mutation caused aggressive outbursts," does momentarily baffle my moral judgments, but the bafflement is unstable. Faced with serious acts of violence, my feelings of moral judgment are surprisingly durable. Ernie might have been genetically engineered to be violent, but *he still did it.* The men with Brunner syndrome inherited a terrible genetic mutation, but they still did awful things. The raped sister might not care why her brother attacked her with a knife.

Some philosophers, the free will and moral responsibility skeptics, argue that my first intuition is the correct one. See how your confidence that someone is responsible is shakier when you think about them being *caused* to behave in that way? They continue: Why would it make a difference, to the person affected by a genetic mutation, whether those causes arose through natural chance versus deliberate engineering? The manipulated agent cannot be responsible.

Others, hard-liners, argue that my second intuition is the correct one. They point to the manipulated agent—the zygote created by the evil Diana, the person genetically engineered by an evil CRISPR scientist, the man fated to violence because he inherited the mutated version of his mother's *MAOA* gene—and insist that they are responsible for themselves, regardless of how they came to be. The philosopher Harry Frankfurt:

> A manipulator may succeed, through his interventions, in providing a person not merely with particular feelings and thoughts but with a new character. That person is then morally responsible for the choices and the conduct to which having this character leads. We are inevitably fashioned and sustained, after all, by circumstances over which we have no control. . . . It is irrelevant whether those causes are operating by virtue of the natural forces that shape our environment or whether they operate through the deliberately manipulative designs of other human agents. We are the sorts of persons we are; and it is what we are, rather than the history of our development, that counts.

Let me tell you the plot of one more horror story about moral luck. In "The Lottery," a short story by Shirley Jackson, the residents of a village gather in the town square, as they do every year. The children play in the dirt and pile rocks and talk about their teachers. The men talk about the crops and the rain, about "tractors and taxes." When the titular lottery begins, early enough so that people can get home for supper, the villagers draw slips of paper from a battered black box. The head of each household draws for himself and his wife and his unmarried children. One slip of paper has a black dot on it. No one looks to see if they have drawn the slip with the dot until every household has drawn. The Hutchinsons have the dotted paper. It is returned to the box, and each member of the family must draw again, one slip for each person, even the toddler. A delegate is appointed to help little Dave, who is too young to draw for himself. This time, Tessie Hutchinson, wife and mother, draws the dot.

Then the villagers throw stones at her.

When "The Lottery" was published in *The New Yorker* in 1948, the magazine received a flurry of mail, and a few canceled subscriptions, from readers who described themselves as outraged, baffled, disgusted. Was it a comment on the Holocaust, on the dangers of conformity, on the banality of evil? Was it even fiction? *The New Yorker* didn't neatly label fiction pieces then, as it does now, trusting that readers could figure out the difference between fact and fantasy themselves. But a postwar audience that in recent years had heard the reports from Dachau and Auschwitz was less certain of the difference. Even now, the story retains its power to disturb.

Tessie Hutchinson's screams express the elemental horror of "The Lottery": "It isn't fair, it isn't right." She receives one of humanity's oldest punishments, stoning, which was historically reserved for what a society deemed the gravest crimes: murder, idolatry, adultery. Her punishment is communal; everyone participates. Even her own toddler is given a few pebbles. But she is not stoned as retribution for some crime. What singles her out as the punished, rather than one of the punishers, is arbitrary chance. She is the victim of bad luck and is brutalized for it. She is a scapegoat.

The story is difficult to forget. How horrible would it be to be part of a society that arbitrarily singles out members of its community and visits brutality upon them? How horrible would it be to be on the receiving end of such arbitrary condemnation? Sitting with these questions excavates a new, dreadful possibility: The story's original readers were right to be confused about whether it was in fact true. Do we not live in a society that singles out members of our community and visits brutality upon them as punishment for their sins?

Yes, you say, but they aren't *arbitrarily* singled out. They are singled out for what they've done, for their misdeeds.

But the line between misdeed and misfortune isn't always so bright as we would like to imagine. Think of the man with Brunner syndrome imprisoned for homicide. He was subject to what philosophers would call "constitutive luck"—the luck of having a certain set of inclinations, temperaments, abilities; of being, psychologically, the sort of person that you are.

Ordinarily, genes do not entirely determine a child's constitution. Even identical twins, who have nearly the exact same genetics, do not end up with the exact same personalities, interests, abilities, inclinations. Instead, a child's constitution depends on how their genetics interact with the physical and social environment, and on "developmental noise," the vagaries and idiosyncrasies that attend any complicated, recursive process. But the complexity of a recipe does not mean that one can ignore the raw ingredients, and a child's genes are a necessary, and unignorable, ingredient in their constitution. Which genes of mine, or of his father's, did my child inherit? This was entirely out of my hands, and out of his. Those genes help shape the sort of person he is, his boldness or inhibition, cruelty or kindness, narcissism or humility, callousness or empathy. And those genes will interact with the environments he will experience in his formative years, environments that I can help shape but cannot entirely control and that he will be largely helpless to change for years to come. Childhood, when we are utterly dependent creatures, is when constitution is constructed.

Some people consider moral luck a fundamentally incoherent concept. They have faith in a strict binary: What is lucky is not moral; what is moral cannot be lucky. Morality is a refuge from fortune. Christians in the first few centuries after Christ, for instance, were very clear on this point. With baptism, they held, we can choose evil over good. We have *autexousia,* which is sometimes translated as "free will" but more literally means "the power to constitute one's own being." There is no such thing as constitutive luck when you can reconstitute yourself, when you can be born again. The early Christians could demand much: celibacy, poverty, asceticism, even martyrdom. But there was an inducement, too, to their view of life. It might be painful, even brutal, but one would never be damned by fortune.

When Christianity became the religion of emperors instead of slaves, however, it was transformed from an ideology that proclaimed humanity's potential for moral freedom to one that emphasized humanity's ineluctable corruption. As Augustine's fateful doctrine of original sin became the new orthodoxy, all humans were seen as afflicted by a type of moral luck: There is nothing you can do to avoid your fallen nature, but because of it, you can do no good and thus deserve eternal torment. Even babies can deserve to go to hell. Elaine Pagels, a historian of religion, describes how this revolution in the Christian understanding of freedom and human nature served the interests of imperial and ecclesiastical authority. Augustine argued that, because even baptized Christians had an inherently corrupted and unfree will, self-government was impossible: It was necessary for the church and the empire to use force, fear, and coercion.

I had given up Christianity by the time I was pregnant with my son, but not its post-Augustinian tragic sensibility. I was less worried about the mythical inheritance of Adam and Eve's sin, though, than about the particular genetic inheritance of my individual child. I was anxious about how his life, and mine, would be affected by, even ruined by, uncontrollable chance.

So far, none of my worst fears has come to pass. My son is impulsive but not cruel. He is concerned when another child is hurt; he wants his teacher to think well of him. One day, as I wait outside for him to be dismissed from school, the father of another boy confides to me that they are thinking about switching to private school. Public school doesn't really know what to do about his son's behavior. His son won't let anyone cut his fingernails. The boy keeps them long so that he can claw and gouge other children if they do something on the playground that displeases him. He sends my son to the nurse's office with a ragged arc of skin missing from his back. The school rewards good behavior with screentime on Friday afternoons, but the boy teases my son for looking forward to his thirty minutes of Minecraft. He asks my son: Why wait all week for a small reward when you could be having *fun* all week, doing what you want?

My son talks about this classmate like he is the serpent in the garden, unpredictable and glamorous and, above all, dangerous. He is just a child. Whatever forces of nature and nurture have conspired to make him delight in harming others are forces over which he has had no more control than Tess Hutchinson had over which slip of paper she drew from the lottery box. Yet in just a few short years, he will be held responsible for whatever sort of man he has become.

The boy is not a character in a horror film. He doesn't have any signs of Brunner syndrome. His is an ordinary sort of cruelty. But once we admit how thin the boundary between bad behavior and bad luck can be, watching him raises the same questions that reading Sophocles and Augustine does. Does he have a choice about the sort of man that he will be? How much choice do any of us have? Is it more tragic or less tragic if he has a choice? These are questions provoked by literature, by theology, and by the science of child development.

But I've jumped ahead. I started telling my story in childbirth. Let's go back to before I became a mother, to how I became a scientist.

Animal

In which I begin experimenting with mice and am astounded to discover that behavior I once deemed sinful can be transformed through science

HOLD OUT YOUR DOMINANT hand. Make an L-shape with your thumb and forefinger, pressing your other three fingers into your palm. Now bend your forefinger at the first joint and angle your thumb up a little higher, a little closer to your hand. The crook you've just made between your thumb and forefinger is perfect for picking up a mouse from its cage. Start at a 60 degree angle. Start from the tail; never face-on. Run your gloved hand up its tiny spine and grip it by the neck, behind the skull.

If you are marking the animal's tail with different colored markers to randomize it to different experimental conditions, don't assign the first mouse you catch in each cage to the same condition, because then the experiment will be confounded by differences in mouse personality. The bold mice, who come out of their corners first, can't all be in the placebo group. Once they've been labeled and assigned, you can start to prep them for surgery.

How did we anesthetize them for surgery? The internet tells me that a behavioral neuroscience lab in the year 2000 might have used an intraperitoneal injection of thiopental or pentobarbital, the same drugs used as sedatives when U.S. prisoners are executed by lethal injection. Or maybe we used a volatile anesthetic: cotton balls soaked in methoxyflurane, pushed to the bottom of a beaker, where the mouse is pushed, too.

But I don't remember that step. One of many things lost to the passage of time. That summer, the last summer of the Clinton years, the last summer of my virginity, exists only in flashes. I am driving a mid-'90s silver Jeep Grand Cherokee with a V8 engine and hood vents. Or maybe I still had the olive-green Oldsmobile? An Alabama boy who is not my boyfriend takes me to see *Coyote Ugly* at the movie theater. I whisper to him in the dark, and I am surprised that his breath changes when my mouth accidentally grazes his ear. I do not yet appreciate the power of desire.

My boyfriend is gone. He is on an Army base learning how to rappel out of a helicopter. His dad flew choppers in Vietnam, and he worries he will never live up to that legacy, never see any real action. He is worried he will spend his career in a peacetime Army. He doesn't have access to an internet connection at Air Assault School, and neither of us has a cellphone, so we write letters to each other on college-ruled notebook paper. In one letter, he affects a sloping longhand and stains the paper brown using a mess hall tea bag. He is pretending he is in a foreign theater, and I am the war bride left at home. Facebook doesn't exist yet, and he can't imagine what he will one day post to it: A photo of him saying goodbye to his bride—not me, thank God, but the girl he dated after we broke up—as he leaves for a yearlong tour of duty. A photo of him piloting a Kiowa Warrior, waving an American flag over Tal Afar, Iraq.

Could that be right, that I didn't have a cellphone then? I don't remember one; I remember only the long-distance code I had to dial to call my parents from my campus landline. I remember eating a handful of miniature chocolate chips from the dining hall ice cream station while walking back to my on-campus apartment. I remember watching a fictional pair of twins, Brandon and Brenda, on the TV show *Beverly Hills, 90210*. Now my first-year college students tell me they envy the Y2K adolescent experience, and I, too, am slightly envious of my past self. I think we are envious of the type of freedom that comes only from disconnection, for time that is entirely one's own. Except for my boyfriend's letters and a few obligatory calls to my parents, I am peaceably cut off from anyone and anything outside the gated confines of my college campus. Such oblivion would be possible only in the deep wilderness now. But in 2000, having finished the first year of my undergraduate degree in psychology, I could retreat from the world merely by staying on campus after finals were over.

I am staying at Furman University in Greenville, South Carolina, for the summer because my father has shocked us all by leaving my mother, and her habitual dissatisfaction with me, the child who most resembles my father, has taken on a new, menacing sharpness. A job on campus, complete with on-campus housing and a tiny stipend, provides me with the perfect excuse to avoid home.

The job is a research assistantship in the lab of a behavioral neuroscience professor who is studying mechanisms of opioid tolerance, withdrawal, and place preference in mice. Tolerance: Why does it take a bigger and bigger hit to get the same relief from pain? Withdrawal: Why does it hurt so much to stop doing something that is hurting us? Place preference: What is so alluring about some cages? To study these questions, we are going to give mice morphine, and then we are going to take it away. But first we are going to tinker with their brains. My only experience with mice, before this summer, was my tween fondness for an afterschool cartoon, *Pinky and the Brain.* Pinky and the Brain are albino mice who have been genetically modified using the Acme Gene Splicer, Bagel Warmer, and Hot Dog Steamer. *They're laboratory mice, their genes have been spliced.* Who knew that was a real thing? But now I find myself manipulating the genomes of laboratory mice to study pain and opiate addiction.

Opiate drugs—morphine, oxycodone, opium, heroin, hydrocodone, fentanyl—make you feel warm, relaxed, euphoric. They blunt the edges of reality so that you feel, at long last, *well.* Opiates work by binding to receptors in your brain that ordinarily respond to "endogenous" opioids, naturally occurring chemicals that are released in the brain in response to activities, like eating and sex and running away from the lion trying to eat you, that promote survival in basic ways. The opioid system encodes liking ("Ooh, that feels good") and interacts with the brain's dopamine reward system, which signals wanting ("Let's do that again"), as well as with a noradrenaline system involved in drowsiness and wakefulness ("Let's go!").

Repeated and prolonged signaling of the opioid system will backfire. The brain, when exposed to an excessively loud opioid signal, down-regulates the receptors available to hear it. As the number of opioid receptors dwindles, a larger and larger amount of drug is necessary to achieve the same effect. When the drug is stopped, the brain is now less responsive to "natural" rewards. Without being down-regulated by the opioid system, excessive amounts of noradrenaline are now released, producing the highly aversive symptoms of withdrawal, a.k.a. dopesickness: diarrhea, sweating, racing pulse, joint pain, tremors, jitteriness, snot, tears, anxiety, sadness, un-wellness. These symptoms of withdrawal motivate continued use. Animals start taking opioids because the drugs feel good, but they keep taking them to avoid feeling wretched.

There are endogenous opioids in the brain, feel-good molecules that produce a runner's high and a postorgasmic glow. There is also an endogenous opioid modulator, called orphanin FQ, which can block the pain-relieving effects of morphine and other drugs. In the summer of 2000, my professor had just won a two-year grant (a rarity, I now realize, for an undergraduate liberal arts college like Furman) to study how orphanin FQ might contribute to opioid tolerance and withdrawal. The ultimate goal of this research was to develop novel painkillers that have fewer side effects and less potential for dependence, and to develop novel treatments for people addicted to opioids and other drugs.

The project was timely. The opioid epidemic had begun, although few people recognized it then. I certainly had no idea. Before the mid-1990s, physicians had been hesitant to prescribe powerful opioid drugs except in the most severe cases of pain or at the end of life. But one pharmaceutical company, Purdue Pharma, set out to change that. They campaigned to convince physicians, and the public, that opioids were a safe and effective treatment for all varieties of pain. It worked. Prescriptions for Oxycontin, Purdue Pharma's extended-release formulation of oxycodone, increased tenfold at the turn of the new millennium. By the time I graduated from college in 2003, the country was awash in pills.

As the opioid epidemic was getting under way, the scientific community began to advocate for a new model of drug abuse. In 1997, the director of the National Institute of Drug Abuse, Alan Leshner, published a paper in *Science* that remains perennially controversial: "Addiction Is a Brain Disease, and It Matters."

Leshner argued against the "common view that drug addicts are weak or bad people, unwilling to lead moral lives and to control their behavior and gratifications. To the contrary, addiction is a chronic, relapsing illness, characterized by compulsive drug seeking and use. The gulf in implications between the 'bad person' view and the 'chronic illness sufferer' view is tremendous." If chronic drug use were better understood as a biological rather than as a moral phenomenon, Leshner argued, then the societal response to it would be compassion and treatment rather than stigma and punishment. (Remember how Julian, the supporter of Pelagius, replied to Augustine? "What is natural cannot be called evil.") Central to Leshner's Pelagian reasoning was neuroscience research, which showed how drugs of abuse affect the brain and how brain changes accompanied people's drug-seeking behaviors. Neuroscience became the vanguard in the war against the war on drugs.

"You are suffering from a chronic disease." Such a dire pronouncement can be a consolation, if your suffering has gone unacknowledged or, worse, has been celebrated as only what you deserve. Here was a scientist, a government official, an authority with power, using science to console rather than to shame, to advocate for mercy rather than to condemn.

I didn't know, when I signed up to work in her lab, that my professor became a neuroscientist because she had spent years suffering from an addiction to alcohol and cocaine. Her adolescent experimentation escalated quickly, because drugs relieved what she eventually could name as existential dread and softened her alienation from other people, who seemed impervious to that dread. She later wrote in her memoir, *Never Enough:* "If others didn't share my horror, or at least consternation, I couldn't understand why not, because we were all subject to the same capricious laws of existence, the same evidence for irrational forces." Even after many decades of sobriety, she continues to have an acute awareness of, and perverse attraction to, the maw of meaninglessness: "Even today, below the persona of nurturing friend, committed partner, determined scientist, and adoring parent is a heartbreaking desire to embrace oblivion." Turning away from oblivion means turning toward the study of those capricious laws of existence, and in particular the laws governing the human brain and its functioning. The former addict became a neuroscientist because she believed that "the few pounds of fatty goop inside my skull were ultimately responsible for the totality of my condition."

In 2000, genetics seemed poised to join neuroscience in unraveling the biology of addiction. In June of that year, President Clinton announced that the first sequencing of the human genome had been completed. His speech was optimistic that genomic knowledge would transform the diagnosis and treatment of hereditary disease: "Today, we are learning the language in which God created life. With this profound new knowledge, humankind is on the verge of gaining immense, new power to heal."

Hereditary disease includes drug addiction, according to many scientists. Adopted children who never live with their biological parents are three times more likely to abuse drugs if their biological parents abused drugs. Identical twins are as similar in their drug abuse as they are in their body mass index (BMI) or coronary artery disease. In Sweden, researchers traced the family trees of over five million people and calculated how often disorders and diseases appeared in people's relatives, ranging from their first-degree relatives (parents, siblings, children) to fifth-degree relatives. Drug abuse was shown to run in families about as strongly as bipolar disorder does. Overall, about 50 percent of the differences between people in their risk for drug addiction can be traced back to the genetic differences between them. Based on "evidence from adoption and twin studies, from genetic inbred strains of rodents, and from induced mutations in mice," Leshner predicted in a 2000 article that the completion of the Human Genome Project would usher in a period of "tremendous progress in . . . the molecular genetics of addiction vulnerability."

At Furman, we are focusing on one gene in particular, *Oprl1*, which codes for the orphanin FQ receptor. Specifically, we are going to inject antisense DNA (a stretch of DNA that is complementary to a DNA sequence that ordinarily codes for a protein) directly into the mouse brain. This, we hope, will block the expression of the gene and thus block synthesis of the receptor protein. The studies require that we be able to make injections directly into the animals' brains when they are awake and moving about. For that, the mice need to have surgery to implant a cannula (a small shunt) into their heads.

The surgery takes place on a small metal apparatus. The jaws of the anesthetized mouse are placed around a bit, and ear clamps are affixed on either side of its head. A tiny dial at the top moves the surgical instrument by millimeters. I adjust it using standardized map coordinates, oriented relative to the sutures of the animal's skull. Science is stereotypically a male endeavor, but this first job in science reminds me of learning to sew. I make tiny movements with a tiny needle, paying close attention to a seam line.

The first few animals are practice. I place the cannula and then inject blue dye through it. My professor takes the first cannulated mouse, still anesthetized, and quickly snaps its neck. "Cervical displacement." She cuts off its head, peels out the skull, and then, like cracking a walnut, extracts the brain to inspect visually whether the blue dye is in the right place. No, a walnut is thicker than a mouse skull. Like cracking a peanut, like cracking a shrimp tail. The practice mice that are uncontaminated by anything that might harm a predator go in a Ziploc bag in the freezer: mouse popsicles, food for birds of prey. Here, too, the science seems so feminine. She is teaching me, and the wisdom she is transmitting is a little bloody and ruthlessly practical. Even the blue dye is reminiscent of menstrual pad advertisements. Women have always known how to cope matter-of-factly with the messiness of bodies, and lab work, it turns out, is no exception. When the dye is reliably in the right place, I am ready to work on animals that will be used in actual experiments. Instead of being beheaded, they will be allowed to recover, and then be allowed access to morphine.

Who had a worse fate: the animals who went to sleep and then were killed before they ever woke up, or the animals who were given morphine and then had it taken away? The latter certainly seemed to experience more pain. Opiate withdrawal in animals can be measured in multiple ways: jumping, wet dog shakes, diarrhea, sensitivity to hot temperatures. My job is to count jumps. How often, within a set period, do all four feet leave the ground? To make counting easier, the dopesick mouse can be set on a small platform, from which they will fling themselves, trying to jump out of their own skin. This is my first attempt at the psychologist's dark art of measurement, in which we attach numbers to inner experience via outwardly visible signs. Prometheus is remembered for bringing fire to humans, but when he recounts the sins that have led to his punishment by Zeus, he starts by describing how he brought us "numbering, chief of all the stratagems."

The antisense DNA technology we are using in the new millennium relies on an ancient insight: The behavior of animals can be changed by changing their inherited biology. Long before there was any knowledge of the mechanisms of heredity, humans selectively bred animals to make them act differently. Fighting fish (*Betta splendens*), for instance, have been bred for nearly a millennium in Thailand, where fights between fish play the same role as dogfights or cockfights elsewhere: a male-dominated occasion for gambling, spectacle, and displays of status and dominance. Selected over countless generations for fighting success, male fighting fish can show lethal aggression within moments of meeting each other, and differences in their aggression have been mapped to differences in their genes.

Going in the other direction, a team of geneticists in Russia has spent the last six decades domesticating Siberian silver foxes. In every generation, the tamest foxes are chosen to be bred. Over many generations, the population of foxes has indeed become more docile, licking human hands, wagging their tails as humans approach, allowing themselves to be picked up and petted. They are also cuter, with floppier ears and tinier jaws.

And, of course, there are different strains of mice. The BALB/cJ mouse is, like Pinky, like Brain, an all-white albino mouse with a big brain, big kidneys, and a tendency to develop cancer in later life. The BALB/cJ mouse was bred for use in biomedical research and accidentally became more aggressive in the process. Mice from this strain attack each other more quickly and more often than other strains of mice, and they don't learn well from punishment, continuing to do things even when they are shocked for doing them. (They also show a particular type of immune response, which makes them useful for medical studies despite their difficult behaviors.)

After the discovery of Brunner syndrome in humans, researchers in the mid-1990s genetically altered one-cell mouse embryos to knock out the *Maoa* gene. The mice bit the experimenters and each other more frequently, leaving scabs and missing fur on their rumps and genitals. They attacked intruders. And when exposed to a virgin, sexually nonreceptive female mouse, the mutated mice "grasped" her more often. Subsequently, a spontaneous mutation in *Maoa* was discovered in a laboratory colony of mice. They, too, bit more often, particularly novel male mice who were perceived as intruders to their cages.

Our study of mice and morphine, like studies of *Maoa*-mutated mice, relies on ancient insight but has modern motivations. We don't only want to change the animal's behavior by changing its biology. We want to change the animal's behavior in hopes of learning how to change ourselves. My professor writes in her grant to the National Institutes of Health that "clinical applications of this work might include the development of novel treatment strategies and pharmacotherapies for opiate addicts." This goal, like so much of science, only makes sense in light of evolution. One can hope to develop new ways to change human behavior by studying molecules in mice only if humans are part of the same grand tree of life as the rest of the animal kingdom, different from, but in continuity with, the birds of the air and the beasts of the field and the rodents in their cages.

The poet John Keats wrote in a letter that "axioms in philosophy are not axioms until they are proved upon our pulses." The axioms of evolutionary theory are, for me, proved upon the pulse of rodents, who also have a ventral striatum, who also have an *Oprl1* gene, who also crave, pace, thrash, anguish. Holding a tiny brain in my palm, I begin to grasp the awesomeness of Darwin's heretical view of humanity as descended from the same family tree as all other animals, rather than created sui generis by God. Language, reason, logic, self-consciousness, poetry, fiction, imagination, meaning—all produced by the differences between the brain I hold and the brain I possess, and those differences evolved, bit by bit, gene by gene, over millennia, rather than being divinely imbued in a single moment.

In this way, my summer job is indoctrinating me into a new applied philosophy, one profoundly at odds with the one offered by my church back home. Collierville Bible Church preached that my fallible and animalistic flesh was attached to an immortal and immaterial soul, corrupted by the fall and in need of redeeming grace. Virtue was to be found in transcending my animal nature; vice was allowing my soul to be dragged down by its appetites. As an Evangelical teenager in the early 2000s, I still saw using illegal drugs as a serious vice, surpassed in its turpitude perhaps only by premarital sex. But my work in the lab suggests that avoiding vice and recovering virtue might be found not in transcending one's animal body but in changing it. Behavior, even behavior condemned as *bad,* might be transformed via interventions to the material flesh rather than salvation of the soul. My professor's goal, after all, was to help people addicted to opiate drugs, not by preaching the gospel of Jesus Christ but by contributing to the development of a new drug. (A quarter century later, a drug that binds to the orphanin FQ receptor is at last in clinical trials. Early results suggest that rats who take it reduce their cocaine use.)

This quest to change moralized behavior through biochemistry might seem less far-fetched now, in the post-Ozempic era, than it did in the early 2000s. Now millions of people have witnessed how taking a pharmaceutical compound can radically change behavior that was once condemned as the deadly sin of gluttony. Rather than "pray the weight away," as one Christian diet book urged, people can instead inject a lab-synthesized compound that mimics a hormone produced in the gut. (Not everyone thinks this medical development is good news. A columnist for *The Catholic Register* complains that Ozempic "circumvents denial of the flesh, carrying of the Cross . . . and our need to rely on God to resist temptation.") Ozempic might even be helpful for addiction: Research has found that people with an opioid use disorder who begin taking Ozempic or a similar drug are 40 percent less likely to overdose. The undeniable effectiveness of Ozempic, the development of which was nearly abandoned because pharmaceutical executives considered weight a matter of willpower, raises the same question that working in a neuroscience lab did: What behavior is within the reach of biological intervention?

I didn't consciously realize what a radical philosophical leap I was making when I chose to work in a neuroscience laboratory. But maybe, on another, deeper level, I suspected. When I was growing up, one of my favorite books was Frank Peretti's *This Present Darkness*. Peretti, a subculture superstar who has sold millions of books, basically invented the Evangelical Christian thriller. The plot of *This Present Darkness* centers on a small band of devout Christians who prayerfully defeat a Satanic conspiracy that aims to take over media, academia, and government. And the villainess of the novel, Juleen Langstrat, is a demon-worshiping psychology professor who beguiles a young undergraduate. Within weeks of becoming, technically, an adult, I ran straight to what my childhood religious community taught me was a hotbed of Satanic activity: the psychology department.

How telling that a conspiratorial and conservative Christian mind imagined psychology to be the most un-Christian place on campus. Not physics, destroyer of worlds. Not economics, globalist traitors. Not sociology, high priests and priestesses of wokeness. Not even biology, the original evolutionist heretics.

But psychology. *Of course* psychology, where knowledge of good and evil is sought in the natural world, where we presume to learn about human behavior—about moral behavior—from animals.

In *Paradise Lost*, Satan appears to Eve as a talking serpent—that is, as a creature transgressing the boundary that God established between human and animal. Eve, who has convinced her only family, Adam, to let her go off and work on her own, does not recoil in horror at the serpent's monstrosity. She instead responds with curiosity: "What may this mean? Language of man pronounced / By tongue of brute, and human sense expressed?" The serpent tells her that he gained his ability to reason from a special fruit he has eaten. He praises both the fruit and Eve: "O sacred, wise, and wisdom-giving plant / Mother of science, now I feel thy power / Within me clear, not only to discern / Things in their causes, but to trace the ways / Of highest agents, deemed however wise . . ."

Is the forbidden fruit the mother of science, or is Eve? Either way, Milton associates the introduction of evil into the world with a feminine desire for knowledge—and not just any knowledge, but science that spans both the "causes of things" and the "ways of agents." Eve, as we know, is beguiled by the prospect. In the last moment before she falls, she wonders, "What hinders then / To reach, and feed at once both body and mind?" In one reading of the Genesis fable, the work I am doing in the psychology lab, which transgresses the boundaries between body and mind, thing and agent, brute and sense, is the work of the devil.

There is another reading of Genesis. In *Cain*, a poetic play by Howard Nemerov, God confides in the world's first murderer.

> GOD: Cain, I will tell you a secret.
> CAIN: I am listening.
> GOD: I was the serpent in the Garden.
> CAIN: I can believe that, but nobody else will.

At the dawn of the millennium, my then-boyfriend thought he'd never see combat. The U.S. presidential candidates thought they would have to decide how to spend the country's budget surplus. I thought I'd save sex for marriage. Many scientists thought the newly sequenced human genome would readily reveal its secrets. And a convicted murderer named Jeffrey Landrigan thought he might be able to get out of a death sentence by arguing that his biological inheritance predestined him to violence.

Landrigan's problems emerged early in life. By the time he was eleven years old, he wasn't just hurting other children on the playground. He was also using alcohol and drugs, shoplifting, and burglarizing homes. His parents, who had adopted him in a "closed adoption" that precluded all knowledge of and contact with his biological parents, were mystified. Why was he doing these things?

As a young man, Landrigan fatally stabbed his so-called best friend and was imprisoned in his home state of Oklahoma for the murder. While in prison, Landrigan began to learn about his biological family. Another inmate had met Landrigan's biological father, Darrell Hill, during a previous prison stint and recognized the familial resemblance. Darrell was still in prison himself. After a history of rape and assault, he had ended up on death row in Arkansas for committing a double murder. Darrell's own father, Landrigan's grandfather, had also run afoul of the law: He had reportedly been killed in a shoot-out with police.

While in prison for his first murder, Landrigan repeatedly stabbed another inmate, was convicted of *that* crime, and then escaped. He ended up in Arizona, where he met Chester Dean Dyer, who invited Landrigan to his apartment. The two men drank beer, had sex, and then Landrigan strangled Dyer with a power cable and stabbed him with a screwdriver. Next to the screwdriver, Landrigan left a half-eaten sandwich. He was soon caught and convicted of first-degree murder. At trial, the judge remarked that the murder itself was not "out of the ordinary when one considers first degree murder." (That judge had seen some things.) But she found the murderer himself quite unusual: "Mr. Landrigan appears to be somewhat of an exceptional human being . . . a person who has no scruples and no regard for human life and human beings . . . Mr. Landrigan appears to be an amoral person."

In the language of the American Psychiatric Association's *Diagnostic and Statistical Manual,* Landrigan, who persistently violated social and moral norms, who lied, cheated, fought, conned, threatened, and hurt on purpose, would have been diagnosed with conduct disorder (CD) when he was still a child and teenager, and with antisocial personality disorder (ASPD) as an adult. Not surprisingly, conduct problems are one of the most common reasons that children and adolescents are referred for mental health treatment, with about 3 percent of school-age children qualifying for a CD diagnosis. Boys are diagnosed at about twice the rate of girls. This prevalence is consistent across the countries where CD has been studied, as is the sex ratio. The prevalence of CD is also generally similar across racial and ethnic groups, with apparent differences between groups accounted for by differences in socioeconomic status, concentrated neighborhood poverty, and other known environmental risk factors. (Landrigan identified as white and was raised in a middle-class white family.)

About half of children who meet the diagnostic criteria for CD will, once they turn eighteen, meet the diagnostic criteria for the adult form of the disorder, ASPD. One childhood characteristic that predicts a "life-course persistent" pattern of antisocial behavior, and particularly violent behavior, that stretches into adulthood is "callous-unemotionality," characterized by lack of empathy and remorse. Some children hurt other children and feel bad about it; some hurt other children and think it's fun.

The adult manifestation of extreme callous-unemotionality is psychopathy, which is familiar from fiction and true crime: Cathy in *East of Eden,* Hannibal Lecter in *The Silence of the Lambs,* Patrick Bateman in *American Psycho.* Ted Bundy and Jeffrey Dahmer. These characters, real and imagined, seem like moral monsters, qualitatively different from ordinary people. But psychopathy is better understood not as a rare syndrome, afflicting the few, but as a continuum, like height, stretching through all of humanity. Vanishingly few people are over seven feet tall; even fewer can chat casually about murdering dozens of women, as Ted Bundy did. But there are people we might describe as "tall," and are certainly taller than average, without developing such an extreme phenotype. Just as there are many people who are taller than average, there are many people with more shallow emotions, less fear and anxiety, more muted empathy with others' distress, and less guilt or remorse.

Psychopathy, in turn, is part of what personality researchers call the "dark triad" of personality traits, along with narcissism and Machiavellianism. Narcissism is characterized by an inflated self-perception and an excessive interest in gaining admiration and special treatment from others, while Machiavellianism is characterized by duplicitousness, a tendency to see people as instruments to be manipulated to maximize one's self-interest, and a cynical disregard for moral norms. The dark triad is remarkably consistent across cultures. One study of forty-nine countries found that, around the world, some people are recognizably more cruel, indifferent, vain, self-interested, and manipulative than others.

As pioneering psychopathy researcher Robert Hare noted, "Not all psychopaths are in prison—some are in the boardroom." The *DSM* emphasizes criminalized forms of antisocial behavior, like brandishing a weapon or setting fires, but the dark triad predicts engagement in unethical behaviors that are not necessarily criminalized, like online trolling, emotional abuse of intimate partners, and employee intimidation, as well as corporate crimes like fraud, embezzlement, and corruption. These acts can be every bit as harmful as street crimes, if not more so. Compare, for instance, the consequences of pharmaceutical executives lying to the FDA, doctors, and the public about the addictive potential of prescription opioid drugs, versus the consequences of an adolescent shoplifting to buy those same opioid drugs. Both are antisocial behaviors that harm others for personal gain, but only the latter is reflected in the psychiatric diagnostic system. (And only the latter has ever resulted in prison time.)

Landrigan was never going to be in the boardroom; he would spend nearly his entire adult life in prison. The jury in Landrigan's murder case convicted him of first-degree murder, and the judge sentenced him to death. In numerous postconviction appeals, Landrigan argued that his defense had been inadequate, because his lawyers had not investigated or introduced evidence about the "biological component" of his predisposition to violence. His father was on death row, a biological father whom he had never met and who had contributed nothing to his rearing, and now he was, too. Was this not evidence that his violence had been caused by the forces of heredity?

Landrigan could have been a participant in an adoption study, where scientists measure the resemblance between children and their biological parents, whom they never met, and their adoptive parents, who raised them. Such adoption studies are the human equivalent of animal cross-fostering studies, such as when rat pups are taken from their mother to be licked and groomed by another. Cross-fostering and adoption studies are considered critical tests of whether there is a hereditary component to a behavior: Are children like their parents when the only thing the parent bequeaths the child is half their genome? In the case of the father, not even the prenatal environment is shared between parent and child. Landrigan's life, in its extremity, illustrated a result that is discernable on a much broader scale: Adopted children whose biological fathers have committed a violent crime are indeed more likely to commit a violent crime themselves.

Adoption studies have a similar underlying logic as studies of identical twins reared apart: Do twins resemble each other when the only thing that ties their lives together is the genome they share? Slightly more complicated are studies of twins reared together, which depends on a comparison between identical twins and fraternal twins. For example, in the 1960s, the psychologist Irv Gottesman (my "academic grandfather," a.k.a. the academic mentor of my academic mentor) co-authored a landmark series of studies showing that if one twin develops schizophrenia, their identical twin is about fifty times more likely than would be expected by chance to develop schizophrenia, too. But fraternal twins or full siblings resemble each other less for schizophrenia than do identical twins; half-siblings even less; and so on. Based on these results, Gottesman estimated that about 80 percent of the differences between people in their risk for schizophrenia were ultimately linked to genetic differences between them. Put more simply: The single best predictor of whether a person will develop schizophrenia is whether they have an identical twin with schizophrenia.

But this is true not just for schizophrenia. For the most severe form of child conduct problems, accompanied by callous-unemotionality, identical twins resemble each other nearly as strongly as they do for schizophrenia. For antisocial behaviors in adulthood—aggression, deceit, reckless disregard for others—identical twins resemble each other as strongly as they do for substance use or body size, more than for depression or anxiety.

Twin and adoption studies of antisocial behavior yield remarkably similar results to twin and adoption studies of drug abuse: They point to a hereditary component to antisocial behavior. The similarities do not end there. As Landrigan's appeals wound through the American legal system, and as researchers sought to bring the promises of the Human Genome Project to fruition, it became clear that many of the specific genes implicated in opioid use disorder are not specific to opioids, or to substance use, but are broadly associated with aggressive and antisocial behaviors and with other violations of moral and social norms.

CADM2, for example, is a gene on the third chromosome that codes for a synaptic cell adhesion molecule. "Synaptic," meaning that these proteins are working in the junctions between neurons, the sites where they communicate with each other. "Adhesion," meaning that these proteins work to connect neurons together. *CADM2* is associated with using opiates, smoking cigarettes, smoking pot, drinking alcohol. It is also associated with having sex at an earlier age, dropping out of school, sleeping during the day, and adding salt to your food. That is, with sloth, gluttony, and lust—with pleasure.

Another example is *FOXP2*, which codes for "forkhead box P2," a protein that regulates the activity of other genes in the genome, turning them on or off. *FOXP2* was first implicated as important for human development in a study that, like the Brunner syndrome study, focused on a single family with a very rare, very serious disorder, in this case a disorder of speech and language. Since then, *FOXP2* has been associated with opioid use disorder as well as with cannabis use and symptoms of ADHD.

FOXP2 also popped up in a study of the BALB/cJ mouse, which, if you'll recall, is a mouse model of "pathological aggression." Compared to mice from a gentler strain, aggressive mice have differences in their *Foxp2* gene. As do aggressive people: *FOXP2* variants were associated with fighting, lying, thieving, disobeying rules at home and at school, and being cruel to animals.

The paper on *FOXP2* in mice and humans, though, sounds a note of caution: "It is clear the association of *FOXP2* variation with ASB [antisocial behavior] has limited explanatory value on its own." Such caution is necessary, because the size of the effect of *FOXP2* on antisocial behavior is extremely small. The probability of being aggressive differs by just a fraction of a percentage point if you have one version of the gene versus another. An effect so small that it might seem entirely ignorable.

Except we keep identifying more and more genes that have these tiny effects. *FOXP2* is associated with hurting people physically, and *CADM2* with impulsivity, and *GABRA2* with drinking too much, and *GPR139* with getting in arguments, and *REV3L* and *WDPCP* and *FURIN* and *ZKSCAN5* and *SMG6* and *NEGR1* and *SCAI* and *SNTG1* and *NCAM1 ASCC3 BPTF SPG7*

ZIC4 CUL3 RANBP17 MAPT NRAP MCHR2 ERAP2 PACSIN3 ICK CCDC88B XKR6 ALMS1 HS6ST3 TMEM110 TMEM163 STK32C IGSF11 SDK1 UTRN AFF3 ZNF75A . . .

In this way, modern genomic studies deliver results that defy circa-2000 predictions, which anticipated finding a few genes with relatively large effects. Instead, the results echo the ancient Greek paradox of the heap. A single grain of sand does not make a heap of sand. Add a second grain: still no heap. Add a third. Add a fourth. . . . Add the nine-hundred thousandth. Add the millionth. Each grain makes an infinitesimally small contribution. Each grain of sand seems like it couldn't possibly make a difference. But as Clov says in Samuel Beckett's play *Endgame*: "Grain upon grain, one by one, and one day, suddenly, there's a heap, a little heap, the impossible heap." Variant upon variant, one by one, and one day, suddenly, there's an impossible heap of genetic difference.

(Clov's next line: "I can't be punished anymore.")

Sometimes, we try to measure the size of a person's genetic heap by adding up all the information we have about as many of their genes as we know about. The results of such analyses land in an uncomfortable liminal space between the overwhelmingly large effect of the Brunner syndrome *MAOA* mutation and the single-grain effect of the *FOXP2* variant. That is, when we jointly consider all the genetic variants that we currently know about, we end up with correlations between genes and behavior that are just big enough to be unignorable but not nearly so large as to be deterministic. These "polygenic" correlations are about as large as the correlations we observe with environmental risk factors for conduct problems: Children develop more conduct problems, on average, when they grow up in poverty, when they are exposed to lead in their water, soil, and homes. When they are bullied by their peers and abused by their parents, when they've been traumatized and neglected. When their mothers smoked cigarettes while they were in the womb, when their fathers were absent. And when they have inherited lots and lots of genetic variants, each of which makes a tiny difference, which together sum to a larger difference.

In the late twentieth century, an American academic proposed a dichotomy: On the one hand, there was "bad" behavior, failures to exercise self-control. On the other hand, there were behaviors, like drug addiction, which could be understood biologically. In the fourth century, a British monk proposed the same dichotomy: On the one hand, there was sin; on the other hand, there were natural events. But if we trace how individual genetic variants are related to drug addiction, we end up identifying genes that are associated with multiple varieties of "bad" behavior, with lying, cheating, fighting, violence. If we trace how risk is transmitted within families, serious antisocial behavior appears as "hereditary" as drug addiction or schizophrenia. The boundary between behaviors that invite moral condemnation and behaviors that can be modeled in animals and mapped to the genome dissolves under scrutiny. Scientific results refuse to abide by the Pelagian antithesis.

These results can provoke a flight into certainty, which preserves a binary between the moral (behaviors that are good or bad, worthy of praise or blame, matters of willpower and virtue) versus the biological (behaviors that are the product of our brains, neurons, cells, genes, and might one day be manipulable by physical intervention) but categorizes all behavior on one side or the other. If even violence is heritable, then heritability be damned: We can ignore the body and treat each other as pure souls; science cannot console us. Or, if even violence is heritable, then morality be damned: We are animals the same as any other, and right and wrong are as imaginary as a talking snake. We will consider both responses more closely, but for now, I invite you not to rush to certainty. Consider the skull in your palm; the fatty goop between your ears; the half-eaten sandwich left next to a screwdriver. Sit with the contradiction of being an animal with a soul.

Landrigan's adopted relatives agreed with his assertion that his genetic inheritance played a definitive role in his life. In a *60 Minutes II* interview with Dan Rather, Landrigan's adoptive sister, who had spent her childhood watching her brother get more aggressive with every passing year, described Landrigan as if he were a character in a Greek play: "I personally think that the day my brother was born, his fate was probably sealed, and that unfortunately, there weren't ways to stop that. . . . And it's a tragedy." Landrigan's biological father, Darrell, similarly expressed no doubts about the role of heredity in his son's murderous behavior: "I don't think there could be any doubt in anybody's mind that he's following his destiny. . . . When he was conceived, that what I was, he became. I believe that."

Landrigan's appeals were unsuccessful. Again and again, the courts found that, even *if* Landrigan could have shown that his behavior had a biological component, that biological story would not have mitigated his punishment—and might have been considered aggravating. In 2001, the Ninth Circuit Court of Appeals wrote, "It is highly doubtful that the sentencing court would have been moved by information that Landrigan was a remorseless, violent killer because he was genetically programmed to be violent, as shown by the fact that he comes from a family of violent people, who are killers also." In 2007, the Supreme Court agreed, quoting the lower court's decision. Inheritance, the court thought, could only make someone more damnable, a line of thought that runs counter to how genetic discoveries have been used to deflect moral condemnation for other behaviors, like drug addiction and body weight. In the pages to come, we will explore the reasons why biological claims, like Landrigan's, might sometimes seem to magnify someone's blameworthiness.

Jeffrey Landrigan made legal headlines one more time before he died, when supply chain issues made it impossible for the state of Arizona to obtain sodium thiopental, a barbiturate used in lethal injections, from a supplier approved by the U.S. Food and Drug Administration. Landrigan's execution was temporarily stopped by a judge, who ruled that obtaining the lethal injection drug from an unapproved and undisclosed source risked violating Landrigan's Eighth Amendment rights against cruel and unusual punishment. The conservative members of the Supreme Court disagreed, and the stay of execution was overturned. Landrigan died by lethal injection in 2010. He was predeceased by Darrell Hill, who, after a quarter century awaiting execution in Arkansas, died of natural causes.

Landrigan was a dangerous and repellent man with a highly unusual family history, an anomaly even among his fellow Cains. The extremity of his behavior, and of his father's, might tempt us to view the question he asked the court as relevant only to outlying cases: If a person has inherited genetic risks that increased their likelihood of engaging in callous, impulsive, lethal aggression, might that mitigate the extent to which they are deserving of blame and punishment, even if it does not entirely absolve them? But this question is not posed only by murderers, and it is not posed only in courts of law. As we will see, "Might my physical inheritance be the reason why?" is a question that has been asked by Christian believers for millennia, and a question that pervades the science headlines today.

Choice

*In which I start graduate school and stop attending church—
an apostasy both volitional and unwitting—and reflect on
how the lives of identical twins embody the tension between freedom and being caused*

THE UNIVERSITY OF VIRGINIA'S campus in Charlottesville is known for Thomas Jefferson's Rotunda, a Palladian temple to reason. University campuses were traditionally centered around a church, but Jefferson, the author of Virginia's statute of religious freedom, centered his academical village around a library. The Rotunda was constructed from nearly one million bricks, which were manufactured by enslaved men, rented by the university from nearby landowners. It was designed to represent, in Jefferson's words, "the illimitable freedom of the human mind."

My office at the University of Virginia was a janitorial closet. Gilmer Hall had none of the Rotunda's elegance, symmetry, or proportion. It was a dim warren of windowless rooms and narrow corridors, clad in concrete. I don't know which grad student first thought to cram a small desk into a corner closet containing buckets and mops, but I'm sure they were drawn to its tiny, improbable window, a rarity in a department that seemed to consider natural light unnecessary for mood and productivity.

The lab I joined at UVA, like my undergraduate lab, studied genetic influences on behavior, including on substance use and addiction. But now the focus was on people instead of mice. No more latex gloves; no more handling rodents. In fact, science no longer involved confronting any animal flesh, not even passing contact with the humans whose behavior I was analyzing. Someone else, years or even decades ago, had weighed and measured the people whose data we used in our research, and I now encountered my subjects only as disembodied entries in vast spreadsheets of data. I spent long hours alone in my closet-office learning how to write code for statistical analyses, to spot trends and patterns, to distinguish signal from noise. I was immersed in a world of matrix algebra, eigenvalues, collinearity, homoscedasticity, factor rotation.

I'm making this sound miserable—the janitorial closet, the buggy code. Not to mention the poverty. I couldn't afford internet; I couldn't afford to eat lunch every day. I lived on a diet of Cinnamon Toast Crunch and Mountain Dew and Taco Bell. I qualified for indigent care assistance from the UVA hospital when I needed an emergency appendectomy, put the rest on a credit card I didn't pay off until I married a man with a trust fund. But I was as content as I've ever been. My life was pared to essentials. The monasticism of graduate school relieved me of myriad daily crises of choice.

Four days a week, I worked on my research. One day a week, I worked at my clinical practicum, an internship designed to teach the basic skills of being a therapist: asking questions, listening actively, formulating diagnoses. My first practicum was in the departmental clinic, where the patients were primarily undergraduates with straightforward anxiety and depression symptoms, carefully screened to avoid any major psychopathology. No psychosis, no suicidality. Nothing too messy or too intractable. Our sessions were conducted in a room with a two-way mirror, surveilled by the clinic supervisor, who could call me during the session on the therapy room landline if I said anything exceptionally gauche or unhelpful.

This combination of training in basic research and training in clinical practice is the hallmark of what is called the "Boulder model" of graduate study in clinical psychology. The goal of the Boulder model is to produce clinician-scientists who can apply the results of objective scientific research to make sense of the subjective experiences of their patients and who can apply first-hand experience with the subjective experiences of patients to design and interpret their research. Science and clinical practice are supposed to be mutually informative.

But I found switching between the two modes of understanding people jerky, disorienting. Most days, I analyzed human behavior as the left-hand side of a regression equation, $\hat{y}$. Human action was the outcome, the response, the dependent variable, the consequence of a web of causal variables. Individuality was relegated to the error term of a statistical formula. Some days, however, I sat with another person and attempted to see the world as they saw it, in all its radical, idiosyncratic subjectivity. I worked to learn and implement therapeutic interventions that assume people have the potential to think differently, behave differently, relate differently, *choose* differently. The challenge set forth by the Boulder model is to integrate knowledge generated from two perspectives: humans as consequences of outside forces, humans as agentic creators of their subjective realities.

In retrospect, I'm not surprised it felt so difficult to knit the two halves of my intellectual and professional training together, the data analyst and the therapist. I now recognize the extraordinary ambition of clinical psychology as a discipline. In his book *The View from Nowhere,* the philosopher Thomas Nagel (the same one who speculated about bat consciousness) articulated the core challenge: "The essential source of the problem is a view of persons and their actions as part of the order of nature. . . . That conception, if pressed, leads to the feeling that we are not agents at all, that we are helpless and not responsible for what we do. Against this judgment the inner view of the agent rebels."

Nagel was expressing in modern academic language what Milton, in *Paradise Lost,* had already put in seventeenth-century poetic form. There, the demons in Hell were the ones doomed, like graduate students, to sit apart and reason endlessly about "Providence, Foreknowledge, Will, and Fate, / Fixt Fate, free will, foreknowledge absolute, / And found no end, in wandring mazes lost."

Ted Chiang's sci-fi novella "Anxiety Is the Dizziness of Freedom" imagines a near future where engineers have invented "prisms," quantum devices that split off two divergent timelines in the universe. At the exact moment when two timelines diverge, the difference between them can be measured only at the level of atoms. But scientists are astounded to learn how much of an effect this infinitesimally small difference ultimately plays in shaping world affairs. Within months, the weather in one timeline is essentially uncorrelated with weather in another. Within a few years, the population on earth is substantially different, because different babies have been conceived.

Meanwhile, ordinary people are largely uninterested in what causes the weather. They want to know about themselves, or rather about their "paraselves," the versions of themselves who exist in other timelines. Paraselves were all once the same self, but their lives have diverged, branching off into new variants with the activation of every prism.

A main character in the story, Dana, is a therapist who specializes in treating patients who have become obsessed with prisms and what they might reveal about the lives the patients are not living. One of Dana's patients has punctured his boss's car tires in a moment of rage. When he learns that none of his paraselves committed the same crime, he feels relieved. His act now seems like a fluke, rather than a reflection of his true moral character.

A second patient wants to find a paraself who said yes to a marriage proposal she declined. If that paraself is also now unhappy in her love life, then maybe she can stop wondering if she made the right choice. How could her choice have caused her current romantic misery if, in an alternate timeline where she chose differently, her outcome was the same?

A third patient, Nat, who is recovering from a drug addiction, is horrified to realize how differently paraselves can behave in different timelines. If there is at least one timeline where one of her paraselves is behaving loathsomely, maybe even murderously, does that multiplicity render her moral choices in this timeline meaningless? She is not seeking to be relieved of her moral responsibility, nor does she consider her paraselves' good behavior reassuring about her own character. She instead begins to fear that all her choices, good or bad, are the product of fluke chance, like billiard balls that just happened to zig rather than zag. It's an anxiety against which she rebels: "I want to know whether my decisions matter!"

In the real world, scientists do not have prisms. We can, however, observe pairs of paraself-like people, who were once the same self but whose timelines diverged, and who are now living different lives—identical twins. For the first two to thirteen days of its existence, the zygote that will later become identical twins is a single organism, conceived from one sperm and one egg, and then, for reasons that are still not entirely clear, the singular becomes plural. At the exact moment of their divergence, the differences between identical twins are imperceptible, measurable only at the level of molecules. But in that moment, their timelines fork. And the emergent differences between twins, or lack thereof, can be as disturbing to our ordinary sense of agency and responsibility, and as open to interpretation, as the differences among the fictional paraselves.

The vast spreadsheets I pored over in my tiny office in graduate school were populated with data from identical twins. (And from fraternal twins, who are conceived from two different eggs and two different sperm.) Our lab at UVA was just one of several labs from around the world that focused on analyzing data from twins, and there was an enormous amount of data to be analyzed. A review paper in 2015 attempted to curate data from every twin study conducted in the past fifty years and found 2,748 scientific papers on 17,804 different life outcomes measured in 14,558,903 pairs of twins. Many psychological ideas now widely accepted as common sense—that autism isn't caused by bad mothering, that people differ in how easily they gain weight for reasons other than willpower, that serious mental illnesses like schizophrenia and bipolar disorder run in families because of genetics—were first advanced on the strength of data from these twin studies.

Despite their popularity, twin studies were, and are, a controversial scientific method. Part of the controversy about twin studies is guilt by terrible association. Nazis were obsessed with twins, performing gruesome and lethal experiments on thousands of pairs who were selected from concentration camps. Far-right reactionaries still twist alleged findings from twin studies to advance eugenic ideas. And part of the controversy is genuinely scientific. Twin studies make numerous statistical assumptions, and not everyone is convinced those assumptions are reasonable, even after several decades of debate.

Within the field of behavior genetics, my lab at UVA was known for its idiosyncratic approach to twin studies. We were generally less interested in the bottomless question of nature versus nurture, in trying to quantify exactly how much twins' similarity was due to a shared DNA sequence versus, say, a shared prenatal environment or shared early attachment experiences. Instead, like Chiang's characters sleuthing out paraselves whose lives have diverged in branching timelines, we were most interested in finding twins whose lives had turned out unalike in some notable way.

For example, my lab mate at UVA, Mary Waldron, wrote her dissertation on pairs of identical twin girls in which one sister had a baby as a teenager and the other didn't. What could we learn about the causes, and consequences, of teenage motherhood from examining these "discordant" twin pairs? Waldron discovered that the identical twin who became a teenage mother was more likely to have been sexually abused when she was a child, while her sister was not. And the identical twin who became a teenage mother remained, long after the child was born, more likely to suffer from depression. In this way, twin study data allowed us to identify where people's paths through life separated, never to reunite.

This work was challenging, in part because systematic and enduring differences between identical twins are rarer than many people imagine. The remarkable convergence in twins' life paths was originally demonstrated in the most famous, and infamous, study of identical twins ever conducted: the Minnesota Study of Twins Reared Apart, or MISTRA, which ran from 1979 until

1999 at the University of Minnesota. Famous because some of its results were as mind-bending as science fiction. Infamous because it was partially funded by the Pioneer Fund, a foundation started by Nazi sympathizers. MISTRA's founder, the psychologist Tom Bouchard, identified eighty-one pairs of identical twins (and an additional fifty-six pairs of fraternal twins) who had all been separated in infancy or early childhood and then raised in different homes. Some pairs of twins had intermittent contact with each other as they grew up, but most met each other for the first time in adulthood, spending, on average, thirty years apart. One pair of twins, Dewayne and Paul, celebrated their first birthday together—at seventy years old—in the University of Minnesota laboratory. Like the separated identical triplets that were the subject of the documentary film *Three Identical Strangers*, some twins did not even know they had a twin until contacted by the researchers.

The results of the MISTRA study revolutionized psychologists' understanding of what shaped people's life outcomes. None of the twins spent their school-age years in the same house, and some of the twins were raised in wildly different environments. The most dramatic example was Jack and Oskar, who were born in 1933 to a Romanian Jewish father and a German Catholic mother and separated at six months old. Oskar was raised in Germany by his mother and was required to join the Hitler Youth. Jack was raised by his father in Trinidad, then a British colony. As adults, Oskar worked in a coal mine, while Jack worked on an Israeli kibbutz. But they arrived at the Minnesota airport wearing nearly identical outfits (light-blue shirts with shoulder epaulettes and wire-rimmed glasses) and scored nearly identically on tests of personality. Overall, identical twins reared apart were surprisingly alike in practically every way that researchers could think to measure, and they grew ever more like each other in their cognitive abilities, personalities, and mental health problems as they developed and aged.

For some twins, these results raised destabilizing questions about what, exactly, in their lives was really their choice and what was better understood as a consequence of the inheritance they shared with their twin. And in a meta twist, identical twins resembled each other in how they interpreted their similarities. Another pair of MISTRA twins was Richard and Robert, both fiction writers and college instructors, both religiously observant, both believers in free will. Robert: "If everything we call the will is just genetics and chemistry, then who in the hell are we talking to when we try to remember something that's on the tip of our tongue?" Richard: "How people react to life is determined by their nature, but I don't think nature is biological. I still believe in good and evil, and that there is such a thing as sin. 'Genes' is just the word we use to describe God."

Richard and Robert's similarity in their religious beliefs exemplifies a more general trend. Along with studying personality, cognitive ability, and physical and mental health, Bouchard and his colleagues measured the religious similarity of all the identical twins who had been reared apart. The correlation between twins can range from −1, which means they are perfect opposites, to 0, which means that they are totally independent, to 1, which means they are perfectly alike. For example, in the United States and Europe today, identical twins, whether reared together or apart, are correlated at about 0.9 for how tall they are. Identical twins reared apart are correlated at about 0.6 for their body mass index. And MISTRA found that identical twins reared apart were correlated at 0.55 for religious fundamentalism and religious values. People who began life as the same life ended up being similar (but not exactly identical) in their adult religiousness, nearly as similar as they ended up being for their body weight.

Subsequent studies have examined the correlations between identical twins reared together, rather than apart, and generally found the same pattern: The more genetically similar you are, the more similar you are in your religiousness. Overall, twin studies of religiousness have found that the strength of genetic influence on religiousness goes up from adolescence to adulthood, as people get more autonomy from their family of origin. By the time people reach middle age, step-siblings or adopted siblings are no more similar to each other than two people plucked from the population, even where they are raised in the same childhood home.

What these twin studies of religion tell you is that part of the reason people are different from one another in the time they devote to religious activities and in the intensity of their religious beliefs is that they were born with different genes. Your religious affiliation—that is, whether you identify as Christian or Jewish or Muslim or Wiccan—seems to be entirely environmental, transmitted culturally from one generation to another. There are no genes that make you wear a cross rather than a hijab. But given that you were raised in a particular religious tradition, your likelihood of actively practicing that religion, of fervently believing that tradition's orthodoxy, of having, in the words of the sociologist Max Weber, a "religiously musical" mind or a religiously unmusical one, is influenced by what genes you inherited from your parents.

Several of the twin studies of religiousness were authored by Lindon Eaves, a late-twentieth-century behavioral geneticist who was also an Episcopalian deacon. The Reverend Eaves reflected on the tension between his religion and his science in an interview with *The Washington Post,* posing questions about the relationship between the subjective and objective perspectives that he didn't answer. That maybe have no answer.

> Science deals with humans in an essentially deterministic way. Whether we talk about genes or environment, we're talking about people being caused. And that's a little bit different than when we talk about ourselves as centered, acting, creating beings. How do we reconcile that with the idea that we're carrying around a code that when it unpacks has a strong impact on us? I mean, I have a paper out there that shows there may be some genetic effects on whether we go to church.

> Inspired by Reverend Eaves's twin studies, my PhD dissertation was "A Behavioral Genetic Investigation of Religiosity and Adolescent Problem Behavior." The premise of the study was to follow siblings who diverged in their religious involvement as they moved from adolescence into adulthood. I tested whether these siblings—one faithful, one apostate—similarly diverged in their alcohol use, drug use, delinquency, and violence. This topic was obviously me-search: How *do* the prodigal daughters of religious families fare in their (im)morality? I turned to the numbers to understand the self.

In my third year of graduate school, it is time to leave the departmental clinic nest. I am placed at the University of Virginia Cancer Center. My role is to shadow a psychologist who treats cancer patients. One of her roles is to lead support groups for people who know that they are going to die. I do little besides watch and listen. I am twenty-two years old. I have little experience with living and none with dying.

A woman in one of our support groups has ovarian cancer, and her prognosis is very poor. She talks often about how her treatments destroyed her sense of taste and smell, and how this robbed her of the pleasure she used to get from cooking for her husband and from eating with him. She and the other women in the group, who are in similarly dire medical straits, don't seem to mind my lack of therapeutic expertise or relevant life experience. I am young; I remind them of their children and grandchildren. Except that I am willing to listen to them talk about the pragmatic details of dying, and they don't have to worry about me being okay while they do it. We practice the conversations they want to have with their spouses and children about wills, advance directives, funerals. My job, I realize, is simply to be there while they practice getting the words out. "I am putting your sister in charge of the estate, not you." "I want to be cremated instead." "I've always hated that priest." "Please don't let me die in this hospital."

While I am working at the Cancer Center, I am also an active member of a large church in Charlottesville. I am passing the peace. I am part of a small group. I am trying to save myself (again) for marriage. I am trying to find people to "do life with." I am *trying*. I am doing everything I know how to do to will my relationship with Jesus to continue.

One Sunday, the church program contains a pink insert. It avoids mentioning any specific candidate, any specific political party, any specific piece of legislation. It merely reiterates the church's commitment to the "sanctity" of marriage between a man and a woman. Around that time, an amendment to the Virginia constitution was wending its way through the legislative process: "Only a union between one man and one woman may be a marriage valid in or recognized by this Commonwealth and its political subdivisions." In Virginia, amendments to the constitution must be passed twice, by two different legislatures, before being sent to voters for approval. The long process gives churches multiple opportunities to instruct their members on what the "godly" stance on the issue is.

I read the insert. I stand up, sit down, kneel down, stand up, sit down. "Peace be with you." "And also with you." I confess my sins: "We have left undone those things which we ought to have done, and we have done those things which we ought not to have done."

Then I happen to see, sitting a few rows in front of me, the woman from my cancer support group, the one who cannot smell or taste. The juxtaposition of person and place snaps something inside me. She is *dying*, and she is with her husband, the husband she wants to cook for, the husband she wants to eat dinner with, and here we all are together, reading a pink pamphlet saying it is good, actually, for some people to be denied the comforts and intimacies of the marital relationship.

I couldn't articulate it then; I can barely articulate it now, why that was the moment I simply couldn't stomach being there anymore. I find myself standing up, one more time, out of sync now with the congregation, and leaving the church in the middle of the service. I don't return.

In the early 1980s, neuroscientists hooked people up to EEG equipment that measured their brain waves and asked them to watch a clock while randomly ("on a whim") deciding when to move their fingers. The participants recorded when they felt aware of the intention to move. The blockbuster result was that spikes of brain activity in the motor planning area, which predicted that people would move their fingers, preceded people's conscious awareness of the intention to move by a fraction of a second. Their brains were unconsciously preparing to move *before* their minds were consciously aware of intending to move.

This result is arresting, because ordinarily we strongly perceive ourselves as conscious intenders of our actions. But such a lag—initiation first, awareness second—can be observed directly in very young children. I have seen this happen in all three of my babies: They hit a hanging rattle with their hand and then startle, stare at their hand. They are surprised that their hand moved, surprised that something emanating from their own body could affect the world. They behold their hand as if it is a foreign object, imbued with mysterious power. When Nietzsche says, "Be sure of this: you are being done at every moment! Mankind has at all times mistaken the active for the passive: it is its eternal grammatical blunder," I think of babies. Infants often appear surprised at their actions, as if they are "being done" rather than doing.

The developmental psychologist Jean Piaget, whose theories of child development were informed by close observation of his own children, called this the sensorimotor stage of development. Infants do things, and observe themselves doing things, and observe the consequences of doing things, until the predictability of cause and effect eventually allows them to invert the temporal relationship between acting and observing, or, at least, their perception of that relationship. As we grow, the observing "I," the self-aware ego, increasingly perceives itself as intending to act before the action happens, rather than being caught off guard, its awareness catching up to action only after the fact. Moreover, that intention is increasingly perceived as having causal power. Adulthood feels active rather than passive. Most of the time.

One of the reasons the famous finger-moving study is contested is because the "decision" that participants were asked to make was so trifling. They were asked to do an inconsequential act for no reason at all. A later study used a similar experimental setup, but people's decision to move was, on some trials, more consequential, determining which charity of two would get a one thousand dollar donation. In consequential decisions, brain activity in the motor planning area was no longer apparent prior to people's conscious awareness of the intention to act. This newer result can be interpreted as a reassurance: For big decisions, at least, we need not worry that we are "being done," rather than doing. This study suggests that, for important decisions, the imperious "I" is still firmly in charge.

But I know I am not alone in having had an experience like the moment when I fled my church: Watching yourself move, finding yourself having chosen, starting to act before realizing what you've started.

I left the church nearly twenty years ago, and writing about it now, I still feel a stab of grief. Growing up, *church* was another word for home. For community, identity, belonging. For love. Leaving the church of your youth is exiling yourself. Leaving the church was among the most consequential decisions of my entire life, and I did it slowly, and then all at once, without consciously realizing what I was doing.

There are, after all, highs and lows in any relationship, moments of connection followed by moments of alienation. The day-to-day variability can mask a slow, dangerous drift in the center of the distribution. The occasional cool day, which gets ever more occasional, lulls you into complacency about how the climate is changing; you can ignore the change until, one day, you realize the ice has melted beneath your feet. The awareness of what is happening, the awareness of what one is doing, can lag behind the doing itself, sometimes by milliseconds, sometimes by years. We do more than we understand.

Some of my friends and family who remained in the church still judge my quitting to be bad. They pray for my soul, because I chose to stray from God's ways like a lost sheep. Others I have gotten to know since then have judged my quitting the church to be good. I chose, in their view, to stop being complicit in the church's homophobia. But here's the thing: "Choice" seems like the wrong word. Does a wind-up toy choose to wind all the way down? Does a runner in an ultramarathon choose to collapse? Sometimes you find yourself done.

In "Cain," the Nemerov poem, Cain is portrayed as someone who wants to find God's favor, but everything he does, unwittingly, exiles him further. The poem opens the same way as the Genesis chapter: God has rejected Cain's offerings of fruit and vegetables, favoring the burnt offerings of his brother Abel's slaughtered lambs. Finally, Cain gets the opportunity to talk to God, and he poses the same question we now ask of science: "Why are things as they are?"

In his reply, God plants the idea of murder in Cain's mind, offering a koan that expresses the power of human will, but also its limits: "Things are as they are / until you decide to change them, / but do not be surprised if afterward things are as they are again." Cain takes up Abel's knife and slays him, and feels himself, for the first time, to be a causal agent in the world. He is now a difference maker. God, too, recognizes Cain's discovery: "You are the discoverer of power."

Except, as Cain contemplates the lifeless body of his brother, as his powers of self-reflection catch up to his powers of action, his feeling of power dissipates. His will caused Abel's death, but his will is a consequence of God's provocation. He has simultaneously discovered his own power and his place in the web of divine providence. His discovery of power, the poem suggests, will lead to civilization, technology, art, and further violence. But it will also alienate him from the God he was, ironically, tragically, trying to draw close to. He rebukes God: "I see it so well, that / You are the master of the will / that works two ways at once, whose action / is its own punishment, the cause / that is its own result."

He vows to reject God, a vow that God initially dismisses. But Cain is resolute: "I do not expect it to be easy."

And then there is silence.

Behavior genetic studies suggest that there are genetic influences on the intensity of religious belief and the regularity of religious behavior. I struggled, and ultimately failed, to avoid the loss of my religion. What, if anything, do objective scientific results and subjective experience have to do with each other? The former is made up of averages, variances, probabilities; the latter of ambivalence, guilt, regret.

A common story we tell, drawing on the science to make sense of the subjective, casts DNA in the role of the antagonist of our lives. I feel ambivalence, guilt, regret about my behavior, which differs from some standard that I hold for myself or from some standard that society has held up as ideal, because "I," the protagonist of my life, have been wrestling with an enemy, and the enemy is my genes. The enemy is my *body*. Scientific findings about the similarity between identical twins are, in this story, interpreted as statements on how strong our opponent is. How overwhelming are the temptations of the flesh? How unlikely are we to succeed in our struggle to master the body? And, perhaps, how much can we be forgiven for failing?

This is one story I could tell you about my leaving the church:

I desperately wanted to stay but somehow couldn't make myself stay, and maybe it's because I didn't have the right genes, the genes for belief, the genes for commitment.

But even as I tell you that story, I remember the limits of genetic prediction. I know that even identical twins differ in their religiousness and their risk for divorce. Do the differences between the branches mean that I can still choose?

I have told my story about religion, but I could tell a similar tale about any number of other behaviors that people regret, and struggle to change, and punish each other for. We all have our confessions. Science finds evidence of genetic influences not just on religiousness but on appetite, weight, physical activity levels; on risk for alcoholism and drug addiction; on depressed mood, irritability, aggression, boredom proneness, and risk tolerance; on extraversion, dutifulness, work ethic; on cruelty, indifference, and compassion. On every form of human behavior. In every crack in human sameness, evidence for the importance of the human genome springs up.

And for nearly every form of human behavior, we also see limits of genetic influence. Except in cases of very rare diseases, like Brunner syndrome, genetic differences between us rarely *determine* behavior. We see differences between identical twins. They differ in their body weight, their mental disorders, their addictions, their personalities, their relationships. Even people who began life as the same life can develop in their own unique, unpredictable way. Even animals who are clones of each other turn out differently. Here in Texas, for example, a couple convinced a lab at Texas A&M to clone Chance, their beloved Brahman bull, who was tame enough to be ridden by David Letterman on late-night TV. The clone, named Second Chance, gored its owners twice.

Genes make a difference; genes are not (usually) destiny.

These dual truths seem to rhyme with our internal conflicts. "You're not imagining it," we hear these studies whisper. "You *weren't* entirely free. Your behavior was constrained in some way. This part of the game of life wasn't entirely up to you." But also: "You're not entirely off the hook. You're not entirely *un*free, either. You still have some control. You are still an agent."

A look at the science headlines reveals how often we tell this story about genes. Newspapers and magazines describe our DNA as our foe, or as our prison, constraining our freedom and opposing our best intentions, but also as something potentially to be overcome. "Can't lose weight? You may have obesity genes to blame." "Six ways to beat your bad genes." "Experts give cunning tips for overcoming the influence of pesky DNA." "Can't quit smoking? Blame your genes." "Yes, you can overcome a genetic predisposition to Type 2 diabetes." "Don't blame kids if they do not enjoy school, study of twins suggests." "Addicted to Facebook? Your genes may be to blame." "Don't blame bad dancers—it's in their genes." "Can we blame procrastination on our genes?" "Healthy lifestyle choices can help women overcome genetic breast cancer risk." "Want to take a nap? Don't feel guilty, it might be in your genes."

Headline after headline, in outlets ranging from frothy tabloids to austere science magazines, all with the same premise, all with the same promise. What is genetic is not free; what is not entirely free is not entirely your fault. Like the Ted Chiang character who hopes to find paraselves who also ended up unhappy in love so she can stop blaming herself for her own misery, we can scan the science headlines searching for twins who are remarkably similar in some way. Both are ways of looking for relief from the pressure of being responsible.

This way of talking about the relationship between genes and behavior is so common we might forget to question it. But here are questions worth asking: Why are genes so commonly associated with unwanted or undesirable behavior, acts that are the focus of regret and ambivalence? Where did genes get their sinister reputation as the enemy of freedom?

One of the most influential books written in the last two millennia was a memoir that described, among other events, the author's abandonment of the religion of his youth, the dissolution of his first marriage, and the significance of differences between twins. Augustine wrote his *Confessions* at the end of the fourth century, as the Christian church was grappling with its relatively recent status as official religion of the Roman empire, rather than as persecuted sect.

Augustine was a convert to Christianity, and before his conversion he had consulted with practitioners of one of the leading competitors to the Christian religion—astrology. In *Confessions*, he made his case for why Christians shouldn't be seduced by the idea that our lives are affected by the movement of heavenly bodies. Look, he said, at the differences between twins, who were born at nearly the exact same time, under nearly the exact same alignment of stars and planets. "Anyone who had examined the one same birth horoscope that applied to Esau and Jacob would have been obliged to foretell the same fate for both of them, whereas in fact their destinies were different." Differences between twins, to Augustine, were yet another piece of evidence that our lives were not merely part of a grand and impersonal cosmic order.

Shortly before he converted to Christianity, Augustine sent away his common-law wife, the mother of his child, so that he could legally wed a wealthy heiress, a separation that left his heart "torn and wounded and trailing blood" and drove him into a fleeting sexual relationship with another woman. Even as he sent away one woman, pledged to marry another, and was sleeping with a third, Augustine debated the merits of a life without sex. His friend Alypius, who was celibate, urged Augustine to renounce sex and marriage. Augustine initially felt this to be impossible. Only after great internal struggle did he decide to convert, forgo marriage, and live in celibacy.

"Decide" makes it sound like his conversion was an exercise in rational choice, but Augustine described it more as an experience of revelation. "A huge storm blew up within me." He heard voices commanding him to take up the Bible: "Pick it up and read, pick it up and read." The first verse he laid eyes on demanded: "Make no provision for the flesh or the gratification of your desires." All at once, clarity and certainty replaced his doubt and ambivalence.

Augustine's complicated pre-conversion sex life had a profound influence on his post-conversion theology. As we all have experienced at one point in our lives, he was tormented by the fact that he could not (or, at least, did not) readily and rationally align his desires and behavior with what he saw as the ideal standard: "I was shackled by weakness of the flesh and was dragging along with me a chain forged of deadly sweetness, fearing to be freed from it." Augustine considered his imperfect command over his own behavior one of the chief miseries of being human. He concluded that the impertinence of the body, which sometimes acts in seeming contradiction to our conscious goals and intentions, must be the result of divine punishment. As he wrote in a later work, *The City of God*, "By the just retribution of the sovereign God whom we refused to be subject to and serve, our flesh, which was subjected to us, now torments us by insubordination." We rebelled against God, so now our bodies rebel against us.

Or, more precisely, Adam and Eve rebelled against God, and

God punished them for their sin with insubordinate flesh. And we have all inherited their sin and so have all inherited their insubordinate flesh. Literally *inherited:* In Augustine's reasoning, original sin was physically transmitted from generation to generation in semen. We cannot help but behave in ways that we will regret, because our freedom of will is compromised by the physical substance of our conception. In Augustine's words: "The flesh lusts against the spirit and the spirit strives against the flesh. I was aligned with both, but more with the desires I approved in myself than those I frowned upon, for in these latter I was not really the agent, since for the most part I was enduring them against my will rather than acting freely."

Do you have desires or behaviors you frown upon? That's the flesh, Augustine says, which you endure, but over which you lack agency. Do you have desires or behaviors you approve of? That's your spirit, which strives against the flesh. Whence did your battle against the flesh come? You inherited it.

Drawing on objective scientific results from behavioral genetics, we can tell a story that makes sense of our subjective experiences of ambivalence, compulsion, regret: Genes partly determine our behavior, and that determinism partly constrains our freedom. Do you see, though, how we have rewritten part of the ancient Christian story with the letters of DNA? We speak of "genes" instead of "flesh"; we speak of "self-control" instead of "spirit." But the new genomic story echoes that old-time religion. We are still describing ourselves as if we were disembodied spirits struggling against a fallen flesh, which constrains our freedom to do as we will.

Sometimes she *feels* real to me, that woman who might have been. The paraself I never had, who took the road not taken. She stayed on the main road of monotheism, didn't wander down the thicket side trail. Thinking about her, I retrace my own steps: When did I branch off? I must admit: No step seemed forced. There was no gun to my head. No extreme circumstances. I am an ordinary woman, with an ordinary psychology, living an ordinary life, making ordinary choices. And when I think about all the tiny choices that constitute my failed relationship with a god, she shimmers more brightly, the woman whom I failed to be, and cannot forget.

Other times, though, she starts to flicker and fade, the woman I might have been. I start to doubt that I was ever *really* free to choose to be her, rather than me. I am haunted by a ghost, and ghosts are just a figment of our imagination. After all, if every step was voluntary, then how did I end up where I would never have volunteered to go?

But this is a familiar lamentation. The disconnect between the power we seem to have over our quotidian choices and our sense of powerlessness over what shaped those choices, our belated awareness of where our choices lead us—this is not just the theme of tragedy. This is the song of middle age.

In "Anxiety Is the Dizziness of Freedom," the therapist, Dana, gently counsels her patients to shift their attention from what might be happening in their paraselves' lives to what is happening in the lives they are currently living. "Consider," she suggests, "whether anything you learn about the other branch would actually be helpful. It could be that nothing you find out about some other branch will change your situation in this branch." To Nat, who wants to know whether her decisions matter, she encourages her to focus on the branching points that lie ahead, rather than the ones already past: "By becoming a better person, you're ensuring that more and more of the branches that split off from this point forward are populated by better versions of you."

The fictional Dana's advice to focus on the current situation, on the future that stretches before you rather than on the alternative ways the past might have played out, reminds me of a lecture by Alan Watts, the twentieth-century English writer and teacher. Watts moved to the United States and was ordained as an Episcopal priest but was then defrocked for adultery, and eventually became one of the first popularizers of ideas from Zen Buddhism and Hinduism for Western audiences. (He remains popular in a subgenre of electronic music that matches melody and beat to snippets of Watts's posh voice.) Watts also encouraged his audience to focus their attention on the present. Even in the Garden of Eden, he points out, the first people did not feel that their behavior was entirely free: Adam blamed Eve for eating the apple, and Eve blamed the serpent. In Watts's retelling, "The serpent didn't make any excuse. He probably winked. Because the serpent—being an angel—was wise enough to know where the present begins. So, you see, if you insist on being moved—being determined—by the past, that's your game. But the fact of the matter is, it all starts right now."

Occasionally I encounter someone who seems to have learned the knack of living simultaneously as if their actions are part of a larger order and as if they are agents whose decisions matter right now. This type of spiritual genius is more common among people who are in recovery from addiction and people who are slowly dying. They confess their powerlessness, their dependence on forces beyond their control, even as they affirm their power to make today different, one day at a time. They seem to grow freer by surrendering to their unfreedom. They straddle two opposing ideas, leaning into one and then the other consistently enough to propel an exhilarating momentum. Watts put it this way: "It's just like riding a bike. It's a balance trick."

That line from Watts is sampled in a song that I heard at the sort of party I would not have been attending if I were living in a branch of my life where I chose to stay in the church. Chose to stay; was caused to stay. As a Christian child, I was taught early to sit with, rather than try to solve, the mystery of an incarnate god. I couldn't understand, but learned to accept, the duality of Christ as both fully god and fully man. Now, as a grown-up, as a scientist, I will spend the rest of my life contemplating, but never solving, the duality of humans as objects of scientific study and consciously self-aware subjects of their own lives. We have wills that work two ways at once; our actions are their own punishment; we are causes that are results. And we must get on with our current lives—the ones happening right now—anyway.

I haven't practiced as a therapist for more than a decade, but I retain a therapist's confidence in one belief: The stories we tell ourselves matter. One story, which casts genes as prisons that constrain our freedom, as antagonists we must overcome, can, like stories of original sin, be deeply comforting. At last! An authority, religious or scientific, is validating our subjective experience: "We" do not always feel in control. We sometimes behold our lives like infants staring at the struck rattle, surprised, even aghast, at what we just did. Sometimes we feel acted upon, rather than like actors. With relief, we give our actions Necessity's name.

But that story has costs as well as comforts. It estranges the self from the flesh: If my genes are my antagonist, then who is the disembodied "I" struggling against them? And it can confound us, when someone else, whom we want to hold responsible for their behavior, lays claim to the same story.

Recall the science headlines, and substitute behaviors you find particularly egregious. "Can't stop stabbing other people?

You have aggression genes to blame." "Finding it hard to quit beating your wife? Blame your genes." "Yes, you can overcome a genetic predisposition to psychopathy." "Don't blame kids if they delight in torturing pets, study of twins suggests." "Addicted to child pornography? Your genes may be to blame." "Don't blame narcissists—it's in their genes." "Can we blame cheating on our genes?" "Want to hurt someone? Don't feel guilty, it might be in your genes." Do they still sound like comforting clickbait?

When confronted by behavior we find particularly blameworthy, we can respond to genetic explanations as the Chorus did to Agamemnon's self-exculpatory invocation of Necessity—with suspicion and condemnation. In the Greek tragedy, Agamemnon tells himself he had no choice but to sacrifice his daughter to appease the gods. But it doesn't matter that the gods narrowed his options; his relief at the conclusion that his choices were unfree is itself portrayed as morally objectionable: "We smelt off him something unclean, something desanctified, something set free to defile." When asked to apply the story of genetic unfreedom to the sins others commit, rather than the ones we are prone to, we realize we can tell a different story. DNA is not always the antagonist flesh against which the spirit battles, limiting its agency and its blameworthiness. It can also, as we will see, become a synecdoche for the corrupted spirit itself, a byword for the reprobate true self.

Essence

In which I question the distinction between nature and nurture and consider how a person's essential goodness or badness—once deemed a matter of their soul's election by God—is now popularly regarded as genetic

ROBERT HARRIS WAS BORN two months premature, after his father, a sergeant in the U.S. Army and a World War II veteran, drunkenly kicked his mother in the abdomen and she started hemorrhaging. Robert was born with signs of fetal alcohol syndrome. His father was later incarcerated for sexually abusing his daughters. His mother was later incarcerated for bank robbery.

Even if he had been adopted at birth, raised by parents who were not abusive alcoholics, a baby with these risk factors would already be expected to struggle. Having biological parents who engage in violence elevates one's own risk for violence. Children who were born early have higher rates of ADHD and struggle more to regulate their negative emotions. Those who were heavily exposed to alcohol in the womb are fussier. They have sleep and feeding problems, are more irritable and more impulsive, have poorer social skills. They are more likely to get in trouble for cheating, lying, fighting. Even rats who are exposed to alcohol in utero and cross-fostered by adoptive rat mothers show impairments in social behavior, including in their own parenting behavior.

MRIs and autopsies of children with fetal alcohol syndrome find that they have smaller brains, on average. Particularly affected is the cerebellum, the "little brain" located at the back of the head. The cerebellum has long been known to be important for motor coordination, but more recently some neuroscientists have reconceptualized the cerebellum as "the mind's quality control center," critical for attention, planning, and decision-making. The cerebellum also emerges as important in genetic studies: Genes associated with antisocial and disinhibited behavior are disproportionately expressed in the cerebellum and disproportionately expressed during the prenatal period. Whether one begins by studying children's toxic exposures or children's genomes, the pathways of risk for antisocial behavior converge—and begin to interact.

Unfortunately for Robert, he was not adopted at birth. Instead, he and his eight siblings were raised by their parents, who seemed to target him especially for abuse and emotional deprivation. One of his sisters recalled, "He'd come up to my mother and just try to rub his little hands on her leg or her arm. He never got touched at all. She'd just push him away or kick him."

Why do parents of multiple children sometimes treat one especially badly, singling them out for negativity or even abuse? Avshalom Caspi and Terrie Moffitt, a husband-and-wife team of psychologists who have long been recognized for their innovative and careful approach to developmental research, carried out a remarkable study in the U.K. in the early 2000s on this question. All the mothers had identical twin children who were seven years old. The researchers found that the twin whose mother treated them with more negativity and less warmth had more antisocial behavior problems. By focusing on the mothers of identical twins, they could be more confident that the children's differential treatment was not driven by the mother responding to genetic differences between her children. After conducting their statistical analyses, they conducted follow-up interviews, asking mothers who felt very differently about their identical offspring to tell them about their beliefs and experiences. They found that sometimes one twin had been much smaller or sicker, or the mother had folk beliefs that twins were always opposites, one masculine and the other feminine, or one dominant and the other submissive. Or the mother's relationship with her partner ended badly, and she believed that one child particularly resembled him and hated the child for it. Even tiny initial differences between the twins could be amplified in a feedback loop by the mother's feelings toward, and treatment of, her children.

Mothers of identical twins can, despite their children's objective similarity, treat them quite differently. Mothers of nontwin children have even more grist for the mill, responding specifically to the unique, genetically influenced characteristics of each child. This is most elegantly demonstrated in the human equivalent of an animal cross-fostering study, a.k.a. an adoption study. Studies of adopted children find evidence of "child effects" on parenting behavior: The child's inherited risk, measured by the characteristics of their birth mother, predicts how their adoptive parents treat them over time. Such child effects are apparent even in closed adoptions where the adoptive parents know little to nothing about the birth parents, and where the child was adopted near birth. One study using this research design found that children whose birth mothers engage in more aggressive and antisocial behavior show more callous-unemotionality in early childhood. This callousness, in turn, predicts their adoptive parents' harsh treatment of them over time.

"Parenting" implies that the parent is the active one, molding a passive child like a potter molds clay. This is absurd. Lumps of clay don't, as all my children have done at one point, fling their bodies to the floor, slam doors, bite and scratch and hit, roll their eyes, scream that they hate you, tell you you're ugly, plot revenge. Treating your child with warmth and affection, while also maintaining clear, firm boundaries, is routinely challenging to even the most well-resourced parents, and some children are more challenging to parent than others. There's a reason that the psychoanalyst Donald Winnicott praised the "good enough" mother. And because most parents raise their own biological children, the children who are most challenging to parent, and whose behavior is perhaps most sensitive to the quality of the parenting they receive, are also the ones most likely to be raised by parents who are themselves impulsive, callous, or mean.

Robert's mother, in these reports, sounds like one of the "monster mothers" in the cruel experiments that Harry Harlow conducted with monkeys in the 1950s, which demonstrated just how crucial the early attachment relationship was for the development of a young primate. Separated from their actual mothers, the young monkeys were given a surrogate "mother," which was just a wire frame attached to a mechanical arm and wrapped in soft terry cloth. The infant monkey would cling to its pitiful surrogate, even when the cloth mother shook it so hard its teeth would rattle, even when the mother unpredictably extruded sharp brass spikes. When introduced later to same-aged monkeys who had been raised normally, with affection and play, the tortured monkeys were rebuffed. Their experiences of isolation and victimization changed how they behaved with others, and others responded to their inexpungible difference by continuing to isolate and victimize them.

The indelible marks left by parental maltreatment can be inscribed on the child's genome. "Methylation" is an epigenetic mark, a sort of chemical tag on the genome. A molecule, called a methyl group, binds to the genome, affecting how genes are "read" by the cell without changing the DNA sequence directly. In animal studies, rat pups who were neglected by their rat mothers showed altered patterns of DNA methylation compared with rat pups who had been generously licked and groomed by their mothers. Specifically, within the first week after birth, they had elevated DNA methylation within the promotor region for the *Nr3c1* gene, which codes for the glucocorticoid receptor gene. This elevation in *Nr3c1* DNA methylation, in turn, heightened the rats' responses to stress, making them more anxious and more aggressive in social groups.

Later work in humans also found evidence for *NR3C1* gene hyper-methylation in children who were victims of abuse or neglect. One study went so far as to obtain brain tissue from people who had died by suicide. Even in this group, united in despair, those who experienced abuse in childhood had alterations in their *NR3C1* DNA methylation in their hippocampus, the seat of memory. Your early experiences with your mother can alter molecular processes in the deepest parts of your brain for the rest of your life.

The psychoanalyst Robert Fairbairn wrote about how children often deal with the intolerable badness of parents and caregivers by assuming the badness themselves: They become riddled with guilt and shame, or they, too, take on the role of the aggressor. Or both. This internalization of badness can feel better for the child, can feel *safer*, despite the suffering it entails, than allowing themselves to recognize fully the cruelty and harm perpetrated by their parent. In Fairbairn's words,

> It is better to be a sinner in a world ruled by God than to live in a world ruled by the Devil. A sinner in a world ruled by God may be bad; but there is a certain sense of security to be de-

rived from the fact that the world around is good. . . . In a world ruled by the Devil the individual may escape the badness of being a sinner; but he is bad because the world around him is bad. Further, he can have no sense of security and no hope of redemption. The only prospect is one of death and destruction.

Robert Harris had a learning disability, as well as a speech disorder, which contributed to his struggling in school. He never got past the eighth grade. At the age of thirteen, he was incarcerated at a youth detention center for stealing a car. There, he was raped multiple times and attempted suicide twice. According to his siblings, Robert was a changed person by the time he was released a few months later. As a child, he had cried when Bambi was shot; as a teenager, he tortured animals, once slashing a pig hundreds of times. He stole another car and spent another three years in juvenile detention, getting out at nineteen. He got married and conceived a son. At the age of twenty-two, he beat someone to death in a fight and was incarcerated again. His wife filed for divorce while he was in prison.

In July 1978, when Robert was twenty-five years old and had been out of prison for just a few months, he and his younger brother, Daniel, were looking for a car to steal to use in a bank robbery when they spotted two teenage boys eating a fast-food lunch in a parking lot. They took the car at gunpoint and allowed the boys to start walking away. Robert then shot both, one of them point-blank in the head. He laughed as he shot them, laughed about the residue of blood and brain and bone on his gun, and laughed as he ate their leftover cheeseburgers. After Daniel threw up from adrenaline and fear, the brothers carried on with their plan to rob a local bank and were caught within hours. One of the arresting officers found out only later that his son was one of Robert's victims. Harris was convicted of murder and sentenced to death. After multiple failed appeals, he was executed in the San Quentin State Prison gas chamber in 1992 at the age of thirty-nine.

That year, Harris's son, who had been renamed Robert Davis after the boy's stepfather, learned of his biological father's identity and his execution for murder. He saw a TV news report while in a juvenile detention center and was shocked to recognize himself in the face on the screen. Just a few days after his release, at the age of seventeen, he robbed a cabdriver at gunpoint and pled guilty to the crime.

In *The Tempest*, Shakespeare chose two nearly-rhyming words to describe the irredeemable character of Caliban, the son of the witch-hag: He was "a devil, a born devil, on whose nature / Nurture can never stick." It is tempting to analyze Robert Harris's life in the same alliterative terms: Were his crimes due to nature or nurture? Was he a devil born or a devil made?

The question presumes that one's birth is separable from one's making, that one's nature can be distinguished from one's nurture, that one's insides are impermeable to the outside. But science, like biography, complicates the distinction between nature and nurture. The parents who emotionally terrorized and neglected Harris also bequeathed him his genes. Prenatal exposure to alcohol affected the same parts of the brain where genes affecting antisocial behavior are expressed. His neurobiological vulnerabilities contributed to problems with emotional regulation, impulsiveness, and awareness of social cues. His social and psychological vulnerabilities, shaped by both his genotype and his early experience, marked him as different, and he was preyed on, first by his parents, then by peers, then by prisons. Such abuse, in turn, affects patterns of gene expression in the brain that perpetuate deviant patterns of stress response and social interaction. To separate "nature" from "nurture" in the development of Robert Harris would require a cleaner separation between inside and outside, self and other. Our bodies shape how other people treat us; other people get into our bodies. The space between nature and nurture, when scrutinized by science, when considering the life of an individual person, can shrink to an extensionless point.

I am not the first to question the distinction between nature and nurture, and I will not be the last. In 1933, the biologist Lancelot Hogben declared the nature-nurture debate to be over: "Genetical science has outgrown the false antithesis between heredity and environment productive of so much futile controversy in the past." In 1958, the psychologist Anne Anastasi—maven of intelligence testing, critic of Nazi race science, namesake of my first backyard chicken—wrote that the "so-called heredity-environment question" was a "dead issue." In 1998, David Lykken, a twin researcher, wrote that asking whether an individual person's behavior "was more due to his genes or to his environment" was "as meaningless as asking whether the area of a rectangle is due more to its length or its width." In 2010, the philosopher of science Evelyn Fox Keller, in her book *The Mirage of a Space Between Nature and Nurture,* concluded, "We now know that the answer is neither nature nor nurture, but both."

Anastasi's 1958 essay was titled "Heredity, Environment, and the Question 'How?'" How, she challenged, do "hereditary factors"—we now call them genes—and specific environmental experiences, such as poverty, abuse, trauma, conflict, alienation, and ordinary unhappiness, combine to shape our thoughts, feelings, behaviors? After finishing my PhD at the University of Virginia, I started my own lab at the University of Texas, and this was the overarching question that animated our research. For every Robert Harris, whose life exemplified the extremes, there are thousands of children who behave impulsively, aggressively, callously, and whose behavior can be understood only in terms of both nature *and* nurture. But which environments made the biggest difference, to which children, when in development? We wanted to go beyond the general platitudes of "nature and nurture" and identify specifics.

In the early 2000s, studying how nature and nurture combine to shape children's development meant finding, recruiting, measuring twins. Some other research groups at the time tried to forgo studying families and instead directly measured a few genetic variants, often called "candidate" genes. But already there were whispers in the scientific community, whispers that would turn into shouts, that focusing on a few candidate genes was not a trustworthy method. Most human behavior isn't influenced by one gene, or even five genes, or ten genes. Most human behavior is influenced by a heap of genetic influence made up of tiny grains of genetic difference. How to measure people's genetic heaps is still a work in progress, and as I was beginning my faculty position, it was a near impossibility. The most reliable way to proceed, if you wanted to study genetics in children, was to study children who had the closest thing to a genetic clone: identical twins.

To find them, we combed through school rosters and filed Freedom of Information Act requests. We showed up at events for "moms of multiples" and sent out newsletters and holiday cards. I carried business cards with me in case I spotted twins at the playground, the grocery store, the trampoline park. Slowly but surely, over the years, hundreds, then thousands, of twins came through my lab. They took tests, mailed us vials of their saliva, lay in an fMRI scanner, gave us pencil-size chunks of their hair. We measured their sensation-seeking, conscientiousness, attitudes about school; their IQs and executive function; their cortisol, testosterone, estrogen; their neighborhood's economic opportunity and family's receipt of public assistance; their relationships with their parents and their parents' relationship with each other; activity in their cingulo-opercular and frontoparietal brain networks. And our results, as long predicted by theory, always found evidence for nature and nurture, working in tandem.

For example: Children genetically predisposed toward faster development in toddlerhood elicit more cognitive stimulation from their parents, which in turn predicts faster language development by kindergarten, a gene→development→environment→development feedback loop that is more pronounced in children whose families have plentiful resources.

For example: Adolescents who inherit genes that push them into puberty at a younger age are more likely to commit minor crimes as adolescents, in part because puberty reorients the brain toward risk and in part because adolescents who look older hang out with older peers.

For example: Adolescents with elevated testosterone levels, as measured in the hair we cut from their scalps, are more aggressive, particularly if they also have low levels of cortisol, whereas high levels of parental monitoring reduces antisocial behavior.

The methods of the child development laboratory do not draw a firm boundary between nature and nurture, born and made, genetics and environment, inside the body and outside. In my lab, genes are another potential difference-maker, but not the only one, or even a privileged one.

Our approach is consistent with philosophers and other scientists who have similarly downplayed the importance of the nature/nurture distinction. Here, for instance, is the philosopher

Daniel Dennett: "What would be so specially bad about genetic determinism? Wouldn't environmental determinism be just as dreadful? . . . Aren't we under just as much of a threat from the dread environment, nasty old Nurture with its insidious indoctrination techniques?" Here is the evolutionary geneticist Richard Dawkins: "Genetic causes and environmental causes are in principle no different from each other. . . . Whatever view one takes on the question of determinism, the insertion of the word 'genetic' is not going to make any difference." He went on to denigrate the idea that genetic influences on behavior were qualitatively distinct from environmental influences as "pernicious rubbish on an almost astrological scale."

The nature/nurture distinction is perennially dismissed by academics, but it is rarely jettisoned in ordinary life, where the work of blaming and absolving is done. People do reason about genetic and environmental influences differently when asked to make judgments about blame and responsibility.

Ask yourself: How do you feel about Robert Harris when told that he was cruelly neglected and abused by his parents, experiences that raise the risk of behaving violently? How do you feel about him when told that he was genetically related to severely antisocial parents, a genetic legacy that raises the risk of behaving violently?

Unlike someone's environmental adversities, their genetic liabilities are not always considered absolving, especially when the domain of judgment moves from the medical to the antisocial. In fact, genetics is often not considered at all. In some studies, participants appear to outright ignore genetic explanations for antisocial behavior, even when investigators explicitly tell them that someone's behavior is due to her genes and give them no other information.

When ordinary people are asked to consider what punishment is appropriate for someone who has committed a serious crime, they are often indifferent to information about the offender's genetics. In one study, the psychiatrist Paul Appelbaum and his colleagues recruited people from across the United States and asked them to read a (hypothetical) legal case involving a man convicted of murder and awaiting sentencing. For example, one case involved a twenty-five-year-old man named John, who gets into a bar fight with a man who talked to his girlfriend and stabs him with a utility knife.

Some of the participants were then given additional information about John's genetics: "The attorney called a psychiatrist who testified that John is biologically inclined to impulsive behavior. The psychiatrist described the results of a genetic test showing that John has a rare form of a gene called IMP7." This information was accompanied by an image showing the specific location of the "*IMP7*" gene. (In reality, no such gene exists, but there are specific locations in the human genome that cause antisocial and impulsive behavior.)

Participants were asked to suggest an appropriate prison sentence, ranging from one to sixty years. (Ask yourself: What sentence would you recommend?) Participants who more strongly believed in the existence of free will recommended longer sentences, whereas participants who had greater scientific knowledge about genetics in general recommended slightly shorter ones. But whether a participant learned about the specific genes of this specific defendant made no difference to sentencing recommendations. The average prison sentence recommended was twenty-four years. Subsequent experiments replicated this result, finding that information about an offender's genetics had no effect on the length of punishment participants found appropriate or their recommendation that an offender receive the death penalty.

Other studies have found that the belief that serious violence can be transmitted genetically from parent to child makes people even more punitive, not less. In a series of experiments led by the psychologist Susan Gelman, people were told about a variety of hypothetical sperm donors who had criminal histories: One stole a pair of jeans, one defaced a priceless artwork, one shot a business competitor after stalking him. Participants were asked how likely it was for a child conceived from the donor's sperm to commit a similar crime, and then were asked how the donor should be punished for their crime, ranging from fines and community service up to the death penalty. Those who rated it as more likely for the child of a sperm donor to resemble their biological father in his crime suggested slightly *more* severe punishment.

Consistent with the experimental evidence, in real-life cases, as with Jeffrey Landrigan, most attempts to introduce genetic evidence about defendants have failed to mitigate the severity of criminal punishment, particularly in sentencing for capital crimes. In 1976, the U.S. Supreme Court established that the death penalty could not be automatically handed out without considering the "mitigating factors stemming from the diverse frailties of humankind." But genetic inheritance is typically not, in the eyes of judges and juries, one of those diverse frailties to be considered when deciding whether a defendant deserves to be killed by the state. Presenting information about an offender's DNA or family history of violence is usually ineffective for reducing punishment, even as the scope of what *is* considered potentially mitigating information about an offender has steadily expanded to include such factors as youth, education, religion, culture, trauma, family background, even neurobiology. A mitigation specialist working with a defendant like Robert Harris today would likely attempt to humanize him by describing all the ways in which his body, brain, and psyche might have been shaped by trauma—but not mention anything about how his response to trauma might have been shaped by the genes he inherited from the very people who traumatized him.

Like a vampire, the nature/nurture distinction refuses to die. Despite the protestations of academic scientists and philosophers, people do reason differently about genetic causes than environmental ones, especially when confronted with exceptionally blameworthy behavior like murder. And like a vampire, the nature/nurture distinction is older than it first appears. It predates the twentieth century, predates any knowledge of genetics, predates Shakespeare's reference to "nature" and "nurture" in *The Tempest*. It stretches back nearly two millennia, at least, to the newly ascendant Christian church as it fended off rival religions.

Lives like Robert Harris's posed a serious dilemma for the early Christian church. Then, as now, it was obvious that we come into this world different from one another—different in our bodies and different in our circumstances. Then, as now, it was obvious that people suffer without having done anything obvious to deserve it; even infants can be born in pain. Then, as now, it was obvious that perpetrators are, so terribly often, also victims. How to reconcile the seeming capriciousness of nature and circumstance with a belief in an all-seeing, all-knowing, all-good God? Human difference, human embodiment, and human suffering posed a serious dilemma for the early Christians, who were motivated to (re)invent a notion of will that shored up their faith.

This was a particularly urgent task because of the challenge posed to the church by Gnosticism, a rival religious movement. Like Christians, the Gnostics believed that the Earth was created by the Heavenly Father of the Jewish scriptures. But they believed that this Father was not the Supreme God but rather a "demiurge," a craftsman. And not a very good one. The Gnostics looked around and concluded that whoever created the world was clearly neither good nor just but was instead more like a shady contractor, or maybe a Trumpesque self-mythologizer, both malevolent and inept. In the Gnostic view, a good God would not have created a world with so much suffering and wickedness. Nor would a good God have made life so *unfair*. Look around you, the Gnostics urged: The suffering! The inequality! The merciless randomness of it all! You think this is the work of a good and all-powerful God?!

In response, the early church fathers needed to develop a case for the Christian view that the Father of the Jewish scriptures was not an evil demiurge but a good and just God. How, then, to explain why people differ so radically in their circumstances and constitutions? How, then, to explain why people suffer so, and seem unable to resist choosing evil instead of good, and sometimes seem caught in the snare of birth and bad luck, without blaming God in his heaven?

Augustine's doctrine of original sin was a clever solution to the church's theodicy problem. As Elaine Pagels summarized in *Adam, Eve, and the Serpent,* "For Augustine, natural and moral evils collapse into one another." This collapse was necessary because, according to Augustine's reasoning, if there were no inherited sin, then infants would be innocent. And if infants who are innocent suffer as they do, then how could God be good? If one abandoned the doctrine of original sin, one might be forced to abandon faith in a just God. Augustine could not countenance the possibility. Like a child of abusive parents, he rationalized that it was better for humanity to be guilty than godless, better to be a sinner in a universe ruled by God than an innocent in a universe ruled by the demiurge, better for our suffering to be deserved than meaningless. He thought people who disagreed with him about original sin would find their faith "shipwrecked upon the misery of infants."

In modern scientific parlance, we might say Augustine came up with a story of epigenetic inheritance, a Lamarckian tale in which acquired characteristics are transmitted to the next generation. "The body keeps the score"—this is the pop science explanation of how trauma is passed down from parent to child; this was Augustine's explanation of how fault is passed down. There was nothing wrong with Adam and Eve originally, but in the garden they acquired a sinfulness, a blameworthiness, that is now passed down to their progeny. What a neat trick: The idea of original sin vindicates God and centers human culpability, while also explaining why some people lead lives marked by tragedy and suffering from the very beginning.

One of Augustine's fiercest critics was Julian of Eclanum, a follower of the British monk Pelagius. Julian disagreed with the doctrine of original sin, and like Pelagius he insisted on a strict separation of nature and morality: "Whatever is natural is shown not to be voluntary. . . . These two, by definition, are opposites, like necessity and will." For Julian, it was necessary to reject the idea that anything physically inherited, anything related to "nature," could substantially affect our moral behavior, to preserve the hope that Christianity held out for moral transformation. He, like earlier generations of Christians, believed that salvation meant being *reborn*, becoming a new creature, remaking one's constitution. Original sin, in his view, was a belief that underestimated the power of salvation. Augustine could not be right about inherited sin if there were to be "hope of humanity's capacity for moral transformation." Julian believed that humans were—or could become, with baptism—"blank slates" in the biblical sense: debts erased, transgressions blotted out, sins forgiven. Opposition to a hereditary predisposition to sin was motivated by belief in the perfectibility of man through Christian baptism.

Nearly two millennia ago, two men were battling to define the orthodoxy of a newly powerful religion. Does what we inherit physically from our ancestors undermine our agency and limit our freedom to choose right from wrong, as Augustine argued? Or could moral behavior never be natural, as Julian insisted? Reading their polemics, the terms of their argument sound strikingly familiar. One side is contending that the body is a site of struggle and a source of fault—and insisting on the hopelessness of people's attempts to change. The other is minimizing the persistent power of the body and asserting that wrong can never be understood in terms of the natural world, in an effort to preserve faith in human perfectibility.

In this way, much of the modern nature/nurture debate was set into motion centuries ago, by two men who were desperate to defend their belief in the goodness of an all-powerful Sky Daddy. The distinction between one's birth and one's making was thought to be demanded by Christian belief.

Not that Christian belief is a relic of the past.

The flight from Austin to Memphis is only the first leg of my planned trip, and the shortest one, but by the time we land I am already exhausted. Boarding a plane as a single mom with a three-year-old and five-year-old in tow is like boarding a plane with leprosy. People look up in alarm, vigilant against the possibility of being near you; then their faces relax in relief as you pass. Even the most exacting disciplinarian has limited control over how a three-year-old will behave when confined for several hours, and I am far from exacting. But the limits of my control do nothing to mitigate my sense of unlimited responsibility for how my children's behavior affects other people. I ply them with toys, books, juice boxes, and snacks; I hold down their little legs so they don't kick the seats in front of them. All the things I'm confessing to you, and yet I still feel compelled to absolve myself on this point: I really do try to keep them from kicking the seats in front of them.

From Memphis, I am going to fly to New York, then to Amsterdam to meet my colleague-turned-lover, and together we are going to fly to Rhodes, Greece, for a two-week vacation. White temples and blue waters, sleeping in past 7 A.M. First, though, I need to bring my children to my mother's house. She has been clamoring for them to stay with her since they were infants; so far, I have resisted. But I have a chance to go on vacation, a *real* vacation, a reprieve from the unrelenting demands of motherhood, from pickup and drop-off and pickup and drop-off, from making dinner and washing dishes and making breakfast and washing dishes and making dinner and washing dishes. I want to go away with such ferocious intensity that I am willing to do almost anything. Even be in my mother's house.

We unload our suitcases in the front hall. I change the children out of their traveling clothes and show them the bin of dress-up costumes in my mother's sunroom. I go to fix myself a glass of water and that's when I see it: an index card taped to the middle of the fridge, right above the ice dispenser. Below it, a picture of White Jesus, with piercing green eyes and curly, light brown hair. The index card has my mother's handwriting, with the top line underlined for emphasis:

> <u>undesirable unwanted</u>
> *I am no longer grieving the loss of "what" Paige could have become—*
> *Her choices—Her responsibility—Her consequences*

Underneath Jesus, my mom has taped up three more notes to herself:

> DIVISION IS BETTER THAN AGREEMENT <u>WITH EVIL</u>
> *The sinner who rejects the shelter (Jesus) has no protection from wrath of <u>God</u>*
> *ReSet my GPS!*

I feel humiliated to be called out by name on my mother's fridge. How many people have seen this scarlet index card: her friends, our family members, the dishwasher repairman? What must they have thought about me?

I am not, however, surprised. "What" I have become is a soon-to-be divorcée; by taking a lover before my divorce is final, "what" I have become is an adulteress. Not a "person who has committed what many Christians consider adultery" but an "adulteress," a phrase that collapses the sin and the sinner, the behavior and the person, into a singular totalizing identity.

When I see the index card on my mother's fridge, I immediately take a photo and text it to a friend. Her response is swift:

> Oh holy FUCK
> What do you even do with that.

> I went to get a glass of water and was like, 'oh hi reason that I've been in and out of therapy for years'

indeed.

> I handled it super maturely by silently ripping it into tiny pieces and saying nothing

> I'm glad you got the photo first
> I'm trying to imagine the kind of life that your mother would have found successful for you. I suspect I would not agree with her image of success.
> That's some pretty wild hair on Jesus

What does it feel like for your identity to be collapsed into just one thing about you? For all your complexity to be flattened into one dimension? It feels wretched. Alienating, and bewildering, and sad.

I cringe to tell you about this familial wound in the same chapter that I discuss the life of someone like Robert Harris, who was so egregiously abused. A refrigerator postcard is not a kick; being privately shamed as an adulteress is not being publicly executed as a murderer. Why, then, do I bring it up? Because, as the American lawyer and human rights activist Bryan Stevenson put it in his book *Just Mercy*, "We are all broken by something. We have all hurt someone and have been hurt. We all share the same condition of brokenness even if our brokenness is not equivalent." If we want to understand *why* someone would do awful things, if we imagine that the answers to *why* might change how we treat someone, then I want to remember that the reasons Robert Harris's life turned out differently than mine—and yours—are differences in degree, not kind.

And the idea that inflicted this specific family wound is an idea much bigger than my individual family. Equating someone's behavior with their identity, categorizing that identity as either good or bad, and then declaring the bad as deserving only wrath: This idea is integral to a Christian worldview. The gospel according to Matthew: "By their fruit you will recognize them. Do people pick grapes from thornbushes, or figs from thistles?" More Matthew: "He will separate the people one from another as a shepherd separates the sheep from the goats." Even more Matthew: "In the time of the harvest I will say to the reapers, gather ye together first the tares, and bind them in bundles to burn them: but gather the wheat into my barn." Paul's letter to the Romans: "Hath not the potter power over the clay, to make out of the same lump one vessel unto honor, and another unto dishonor?" If people's eternal fates are binary—heaven or hell—then their essential selves must also be binary: fig or thistle, sheep or goat, wheat or tare, vessel of honor or vessel of wrath.

As the sociologist Max Weber described, this way of viewing the world, and in particular the fear that one can never *really* know whether one is a goat or a sheep, a saint bound for heaven or a sinner doomed to hell, forever changed society. Protestants uncertain about the fate of their soul engaged in ceaseless industry and in the ceaseless acquisition of wealth, because they were looking for a sign of God's favor. We all now live in the system—capitalism—created by their idea that some humans were inherently goats, doomed to endure the wrath of God.

This idea has also shaped the U.S. criminal justice system. We remain, for instance, the only country in the world to sentence juvenile offenders to life in prison without possibility of parole, a punishment considered fitting when offenders' "conduct shows irretrievable depravity, permanent incorrigibility, or irreparable corruption"—secular terminology for a Christian concept of inherent damnability.

In Christianity, the essence of a person is their immaterial soul, but people need not believe in souls to think essentially. The psychologists Ilan Dar-Nimrod and Steven Heine argue that in our secular, postgenomic age, the idea of "genes" increasingly functions as an essence placeholder in our reasoning about people. One direct-to-consumer genetic-testing kit, for instance, comes in a box proclaiming, "Welcome to you." You will discover your true identity, it is implied, by spitting into a tube.

Even Christians who believe in an immaterial soul now use DNA as a byword for someone's true nature, which can be judged as righteous or unrighteous. The difference between sinners and saints is portrayed as essential, inevitable, ineradicable—and genetic. According to the Gospel Coalition, an organization of Evangelical Protestant pastors in the United States, "The only difference between righteous sheep and unrighteous goats is not what they did or did not do. The difference is not merely in outward characteristics. Sheep and goats have different DNA. The difference is cellular."

This essentialist way of thinking about genes helps explain why people often consider information about someone's genes less absolving of blame than information about their environments: If someone's behavior is "genetic," then it is seen as reflecting who they "really" are, their truest and deepest self, right down to the nucleotides of their DNA. And what they are is a born devil.

This is another modern retelling of an ancient Christian story, except now "genes" are not the flesh against which the soul is struggling. Rather, "genes" are an essence placeholder like the "soul," the epitome of someone's true self. And in this story, genes are not exculpatory, any more than Protestants considered the divine predestination of some souls for hell to be exculpatory. We do not forgive the tares in the field because they could not help but produce their poisonous grains. Their poison is essential to what they are, and what they are demands they be uprooted and burned.

In his paper "Responsibility and the Limits of Evil," the philosopher Gary Watson discusses the life of Robert Harris and considers why knowing about the abuse and trauma he suffered as a child and teenager might change how harshly we think he should be punished. In Watson's view, the details of Harris's childhood history soften our punitive judgments because knowing how he suffered provokes a feeling of sympathy that countervails our moral antipathy. Sympathy, not just pity. Sympathy presupposes some equality of relation; it is feeling *with* someone. When we hear about Harris's brutal childhood, we don't just think about how much he suffered. We are also prompted to consider how *we* would have responded if we had suffered like that. Are you so sure that if you had been treated monstrously when you were a child, you wouldn't have turned into a monster? You, too, could have been morally razed.

Moreover, we can feel sympathy with Harris's moral corruption without necessarily viewing his behavior as the inevitable product of his childhood experience and without necessarily giving up whatever notion of freedom, however minimal, we might have. A belief in determinism, Watson reminds us, is not the only antidote to punitiveness. Jesus's rebuke to the crowd gathered to stone the woman caught in adultery was not "She couldn't help herself." It was "Let he who is without sin cast the first stone." The reminder that softens moral enmity is not that the offender could not have done otherwise. It is that we could have done the same.

Watson's analysis of sympathy also gives us another clue about why genetic causes might affect our judgments of blame differently than environmental causes. When I hear about the childhood history of Robert Harris, I can ponder, "What might I have done in those circumstances?" But the counterfactual for Jeffrey Landrigan, whose upbringing appears to have been unremarkable but whose family tree was highly unusual, provokes a different question: "What might I have done if I had had different parentage altogether, if my genetic constitution were entirely changed?" Such a question can feel unintuitive, bordering on the nonsensical, particularly if the idea of genes is collapsed into the idea of self. Who is this "I" that has an entirely different genotype? While I can imagine my identity being changed even if my genes remained the same, it is substantially more difficult to imagine my identity remaining recognizably mine even if my genes were entirely different. "What if I had had your childhood?" is a question that invites my sympathy for your behavior. But "What if I had had your genome?" is a question that makes sympathy difficult, maybe impossible, because the self who would be doing the sympathizing is threatened with dissolution.

Instead of sympathy, answering such a question requires empathy—feeling *as* another, instead of feeling *with* another. The empathic leap to imagining the subjective consciousness of someone like Landrigan is daunting: What is it like for a person who killed without remorse? What is it like to be cruel and morally indifferent, even as a child? How can we imagine ourselves not just in his shoes but in his skin, body, brain? Imagining his subjectivity might feel as difficult as imagining what it is like for a bat.

But no—Landrigan was human. As was Harris. We must remember: The differences between me and them, between you and them, are differences of degree, not kind.

My only way into empathy, narrow and uncomfortable, is to remember moments when I have been unable to feel affection or compassion for someone vulnerable, or unable to muster the slightest bit of concern for what I "should" do in a situation.

Here is one: It is 4 A.M. and I am wild-eyed with sleep deprivation and postpartum anxiety, looking at my hungry, squalling infant, unable to care enough about her misery in that moment to pick her up and let her continue suckling the life out of me. I stare at her dully as she screams.

Here is another: A woman in a large SUV turns right, across the bike lane, and nearly hits the bike trailer that holds my toddler. I lose conscious awareness; I quite literally black out, with no awareness or memory of what comes next. Not because I've been hit, but because I'm overcome by primal, maternal rage. I return to awareness to find myself leaning into the driver-side window of the offending SUV, screaming obscenities.

Trifling moments, compared to murder. But moments when I felt only numbness, or only rage, emotions that are opposite in intensity but that both sever connection with other people. Moments when I was stranded in my own skin suit, alone. Is this how Landrigan and Harris felt every day, every hour, every moment? If so, their lives were a living hell long before they were imprisoned. Here, I think, Augustine was correct: Living as a body capable of great evil can be punishment in itself.

As I allow myself to imagine his subjective experience, the objective questions about what caused Landrigan's behavior, or Harris's, or yours, or mine, recede. Nature or nurture, mutation or trauma, or none of the above? Regardless of the answer, whether someone seems to be a devil born or made, an occasional sinner or an incorrigible reprobate, I do not want to answer others with the same callousness and cruelty that I condemn in others. "What" they are is *people,* our fellow humans. All of humanity is in the same boat, with only unearned chance—another word is *grace*—separating a person from a life of wrath. There, but for grace, go we. There, but for grace, *are* we.

One danger of genetic essentialism, then, is that it can estrange us from others. Imagining someone as having a qualitatively different essence makes empathic connection more difficult. Dar-Nimrod and Heine warn of another danger: Believing in a single, fallen essence can engender a false hope in a single, redemptive cure. "Essentialist biases allow us to be mesmerized by the siren call of genetic solutions to life's problems," they write. If "an improvable essence" (the genome) "underlies all human challenges and glories," one might be tempted to try to improve it. In other words, essentialism can beget eugenics. Their warning about eugenics echoes the warning in Milton's *Paradise Lost.* Eve, too, was beguiled by the prospect of a single, natural solution to all of life's problems: "Here grows the Cure of all, this Fruit Divine . . ." Eve pursued a final Cure and found death. She was not, as we will see, the last.

Variety

In which I miscarry a wanted pregnancy and, at my middle child's urging, wrestle with what we long for and what we dread when we try to control reproduction

IN THE SUMMER OF 2023, the children and I are in the front yard, in a square patch of grass we've faithfully watered through the long Austin heat. The variety of St. Augustine grass in this patch is not native to Texas; it is not native to anywhere. St. Augustine grass is indeed named for St. Augustine, the man, by way of St. Augustine, the city, which got its name because Catholic conquistadors from Spain spied the land where they founded the city on the saint's feast day. The variety of St. Augustine grass in our lawn is called "Floratam," a portmanteau of the names of two universities, "Florida" and "Texas A&M," because it was cultivated by university scientists who were tinkering with the forces of reproduction. St. Augustine grass forms a dense mat that chokes out competitors, producing a uniform-looking expanse, the classic American dream. But its lushness belies its fragility. A lawn colonized by a single plant can be destroyed by a single misfortune: drought, cinch bugs, fungus.

Our patch of green lawn is surrounded by a low rabbit fence, to keep the baby from wandering off, and beyond that by yellowed stalks of wild grasses and scrabbly mesquite upstarts. The wild grasses are drought-tolerant, adapted to the heavy clay soils of the Blackland Prairie. They provide habitat to pollinators and slow water runoff, but they are inhospitable to a toddler's feet. My third child, my baby with Travis, is now one year old, so we have fenced and watered this patch of St. Augustine grass for her, so she can have a place to play.

"For her," allegedly, but really for me. I want to take her outside, sometimes, without having to be too vigilant. There are mosquitoes out here, and a few pointy sticks, but otherwise the hazards are few. I can sit in an Adirondack chair under our oak tree and watch her and almost relax. But she is less enthusiastic than I am about the front-yard play area. By scrubbing it of danger, we have scrubbed it of interest. She would rather play in the backyard.

There is no St. Augustine grass in the backyard. Instead, spread over nearly an acre, is bastard cabbage and bee balm, bloodberry and black-eyed Susans, Anacacho orchids and Engelmann's daisies, prickly lettuce and widow's tears, dwarf palmetto and wild petunia. And stinging nettle, toothed spurge, giant ragweed. Our property's previous owners had a pony, a miniature horse, and a flock of hens and roosters, who stamped, grazed, and pecked the land to bare dirt. They also shit everywhere, extruding seeds far from their origin. We haven't done much new planting. Instead, we have been watching, waiting, selecting. We see what springs up from the soil and decide whether we want to keep it. Some plants we let grow; others we pull on sight. A plant identification app is now one of my most-used phone features. The ability to diagnose the tiniest green sprouts is valuable when you are trying to uproot things early. I am a scientist with gloved hands, tinkering with the forces of reproduction, but the old-fashioned way, with a spade and a hoe.

The decisions about what to keep and what to uproot are

often more difficult than I imagined they would be. I am tasked with assigning meaning and value to biological variation, but the varieties of life do not conform to neat categories. Heartleaf nettle, for instance, is a native plant that supports mourning cloak and painted lady butterflies. It is an edible herb used for centuries to treat arthritis and inflammation. It seeds on the wind and is difficult to contain, but it will die back with the summer heat. And it has tiny hairs that burn on contact, leaving a rash that can linger for days. We pull most nettles but deliberately keep a few, exceptions that are never allowed to become the rule. Our neighbor across the street is appalled. She keeps chickens just to prevent nettle from growing; she reminds me that the wind doesn't respect property lines. (Neither do deer, which she feeds.) Our individual decisions have communal consequences.

The Wild Iris is a volume of poems by Louise Glück that chronicles a year in the garden. The plants pray to the gardener; like Job in the Bible, they bemoan waiting, suffering, injustice. As Travis and I discuss the nettle, I think of one poem, "Clover," told from the perspective of a trampled weed rebuking the ambivalent garden-God for prizing the four-leaf variety: "by what logic / do you hoard / a single tendril / of something you want / dead?" The four-leaf clover is valued as lucky; clover is devalued as a weed. But one does not go without the other. One *is* the other.

Much of what has sprung up, since we moved in, are plants like heartleaf nettle—hardy, resilient, opportunistic plants, which are commonly found in what my plant identification app calls "disturbed areas," meaning areas disturbed by human existence. They grow in abandoned buildings, cracked sidewalks, highway underpasses, urban riverbeds, and backyards formerly occupied by miniature horses and too many roosters. Such plants can seem wild, in that they were neither deliberately planned nor planted, but their existence is nonetheless closely tied to human culture, to the consequences of human action. Or inaction. One of my father's personal maxims rings in my head every day I spend in the yard: "Not deciding is deciding." Even if I never lifted a hand in the garden again, never planted or watered or pulled a thing, the resulting landscape would still be a human creation.

In "The Queerness of Eve," a series of poems by Emilia Phillips, Eden is imagined as a place like my backyard, full of allegedly undesirable life, overrun by the unintended and difficult to contain, delightful. "The Garden was prolific / in wild invasives." And it is delightful, even to the baby. Especially to the baby. She is getting to know the world. She has an experience-expecting brain that orients toward novelty, uncertainty, variety, risk. She wants dappled shadows and rocky slopes and unfamiliar flowers.

But her growing acquaintance with the world again confronts me with all the ways the world could hurt her. The world *will* hurt her. The front yard, with its patch of highly unnatural nature, is a tiny respite from my unrelenting maternal anxiety. Nettles are valuable. And I want to protect my child from nettles. A biologically diverse landscape is good, and so is my child being spared suffering. These values coexist, always in tension.

On this morning in August, then, we are not in the backyard. We are lolling in the still-green grass, half-heartedly kicking a soccer ball, when my older daughter, who is eight, surprises me by asking: "What is a miscarriage?" The question has no obvious prompt, but she has a habit of asking questions with no obvious prompt. "The vault," Travis calls it. She will hear or read something—a stray remark, a passing reference, a conversation in another room—and give no outward sign that it has made an impression. Then, weeks or months later, when she has made a nest for that knowledge in her mind, she will ask me.

We have already talked this year about birds and bees, penises and vaginas, periods and pregnancy, so she has some background knowledge to understand miscarriage. Before having kids, I imagined that one day I would give "the talk," like my mother did—a single, excruciatingly awkward monologue where the clinical details of how babies are made were swiftly imparted and never spoken of again. But I end up trying a different tack with my own kids. I buy a sex education comic book with cheeky, cheerful illustrations and suggest we read it together, adopting its nonchalant tone. Having pretended to my children that I find nothing embarrassing about the subject, I find myself trapped in a seemingly endless series of conversations about puberty, sex, and reproduction. This open line of communication is, I reassure myself, a good thing overall, but I am faking my blasé demeanor. I am, in fact, embarrassed. I do not want to talk about pubic hair with an eight-year-old. Or miscarriage.

"Miscarriage is when an egg and a sperm have combined to make a baby, but for some reason, the baby can't live in the mommy's uterus anymore. It dies, and it comes out like a period, like blood. Most of the time, a miscarriage happens very early in the pregnancy, so it doesn't look like a baby yet; it's just a bundle of cells. But sometimes it happens later, and sometimes a doctor has to help."

"Have you ever had a miscarriage?"

I tell her yes, I had one miscarriage. Before I was pregnant with the baby who is now toddling around the lawn, I was pregnant with another baby, but then I lost it.

I miscarried early in that pregnancy; there were no complications; the pain was manageable. I was comfortable at home, and my body expelled everything on its own. The bleeding and the cramps stopped after about a week. The most difficult part, in the end, was how little time I had to grieve. My first book had just come out, less than two weeks before my miscarriage. That book launch entailed a punishing schedule of interviews about difficult topics, made more difficult by my roiling hormones and overall feeling of physical vulnerability. That week, I wanted to rot in bed in my red flannel pajama pants and watch *Buffy the Vampire Slayer*, comfort items dating back to high school. Instead, I did a series of podcasts and lectures about genome-wide association studies. The soft animal of my body was packed up with extra-large sanitary pads so I could perform being a brain.

Despite the mess and the grief, I count myself as fortunate. Exactly one month before my miscarriage, on September 1, 2021, the "Texas Heartbeat Act" had gone into effect, banning abortion after about six weeks' gestation, or just two weeks after a missed period. The law authorized antiabortion bounty hunters to sue for ten thousand dollars anyone performing or "facilitating" an illegal abortion. Among its many terrible consequences for reproductive health, the law has changed how doctors treat miscarriages. The womb might be open, or the membranes ruptured; miscarriage might be inevitable and sepsis imminent. But if there are fetal heart tones, then doctors in Texas—and now in many other states with similar bans—cannot perform the surgical procedure to vacate the uterus. Miscarrying women have been sent home with blood filling their shoes, have clawed their bathroom walls in agony, have gone into septic shock and become infertile. When I miscarried, I was not just sad about the end of a much-wanted pregnancy but also anxious about what would happen if I needed medical help that I couldn't get. If we had to leave Texas, should we drive or fly?

Where? Who would watch my older children? How early would I know if something was going seriously wrong? That I didn't end up facing these dilemmas felt like a pardon.

Texas has now banned abortion outright, even in cases of rape and incest, except when an abortion is necessary to save the life of the mother or to prevent "substantial impairment of major bodily function." This ban became effective in 2022 after the U.S. Supreme Court decided in *Dobbs v. Jackson* that Americans did not have a constitutionally protected right to abortion. A week after that Supreme Court decision, I was delivered of my third child by cesarean section and underwent a tubal ligation. After twenty-five years of being a fertile person, I felt nothing but relief at my new sterility. I no longer face the hazard of becoming pregnant in the state of Texas, where another pregnancy could be ruinous or lethal. I do, however, have daughters, one of whom is only a bit younger than the ten-year-old Ohio girl who had to travel out of state for an abortion in June 2022.

I don't tell my daughter any of this. I just tell her that I was pregnant, once, and then I wasn't. That I didn't know if it was a boy or a girl, or why it died, but that there might have been something wrong with its DNA. That I was sad when I lost the baby, but then I got pregnant again, right away, and didn't feel sad anymore. That, no, I'm not afraid it will happen again, because I won't ever get pregnant again. The doctor who cut her baby sister out of my belly also cut my fallopian tubes.

My daughter considers this information for a while, and then she blurts out: "I'm glad that other baby died, because I really like the baby we have."

I have raised my daughter free of the church; we don't even go at Christmas or Easter. Yet I'm still momentarily stunned by her paganism. She doesn't ask, as I would have at her age, whether the baby went to heaven or hell after it died. Such a question would, I think, not even make sense to her. She thinks that heaven and hell are stories, like the stories about Hades and the Underworld in her Percy Jackson story books. To her, the miscarried fetus was a temporary body, not an eternal soul; a potentiality, not a person. Her comment assumes the "it"-ness of a human fetus. What seems important to her is that the people whom she does see as persons got what they wanted. Travis and I wanted a baby, and we got one. She wanted a little sister to play with, and she got one. The baby, she assumes, wants to be alive, and is—and would not have been if I had not miscarried the month before she was conceived. Her family is the people she can see and touch and love; the miscarried fetus is unimportant by comparison. She cannot yet fathom that the state would force her to remain pregnant for the sake of such an "it."

I think about my daughter's comment all the time. When I'm bathing the baby, delighting in the rolls of fat on her arms and legs, or when I'm rubbing noses with her, noticing her amber eyes, I wonder: "Am I glad?" Am I glad my other pregnancy ended? Even now I use a euphemism, "pregnancy ended," to avoid saying what my daughter stated so plainly: The fetus died. There was something alive inside of me, something both human and alien, and it died.

Glad is not quite the right word. I am attached to the baby the life—that I currently have, and so would not choose, if I had the opportunity, to rewind the clock and undo the miscarriage. But I would not be the first woman to feel glad that a fetus died. Studies of women who have abortions find that their most common emotions in the days, weeks, months, years after the abortion are relief and happiness. An analysis of nineteenth-century letters about miscarriage found that some women, "exhausted by a married life of never-ending reproduction," were "relieved, or even overjoyed" to miscarry. They mailed expelled material off to scientists trying to understand reproduction. I might feel differently if my other pregnancy had been more advanced, if I hadn't gotten pregnant again so quickly, if I weren't so enamored of my youngest. But it wasn't, I did, and I am.

I wouldn't have let myself admit it if my daughter hadn't said it first. I am squeamish about admitting that something so precious—a human fetus, a human *life*—was also, to me, something fungible, replaceable, superseded. But I also know: Reproduction involves ruthless trade-offs. Having one baby requires not having had the other one. Having my particular life, which I want, means having had to sacrifice the opportunity for other potential lives. Something flowers only if something else doesn't take root first.

My daughter's intuitive calculus, and her stark way of expressing it, reminds me of the controversial views of the utilitarian philosopher Peter Singer, who has argued, as have others, that fetuses are not persons. It is okay, or even good, if some fetuses don't live, especially unwanted ones or ones who would be born with a disability. In particular, Singer has argued that the lives of people with serious disabilities are inherently worse off, so much so that their lives are best prevented. If a fetus who will become disabled dies, we should be glad. The parents can conceive another child if they want one, as we did. Their lives and the life of their surviving child and the entire community will be better off than if the fetus had lived.

To call Singer's views "controversial" is putting it much too mildly, especially because he extends his logic beyond the womb. Newborns, he argues, are also not people, and letting some newborns die is better than letting them live. In *Should the Baby Live? The Problem of Handicapped Infants,* the 1985 book he wrote with Helga Kuhse, they put their thesis plainly: "We think some infants with severe disabilities should be killed." Even after birth, they argue, newborns are not persons, not in the way you and I are persons. Newborns, because they "cannot see themselves as beings who might or might not have a future, and so cannot have a desire to continue living," haven't "properly" begun life's journey yet, so there is still an opportunity, even after birth, to prevent a life not worth living.

Many who live with quite serious disabilities, however, testify to the wholeness and quality of their lives and reject whatever utilitarian math says they are not worth living. In a remarkable essay for *The New York Times Magazine*, Harriet McBryde Johnson, an attorney, writer, and activist who was born with a genetic neuromuscular disability, characterized Singer as "the man who wants me dead" and compared him to a Nazi. Singer has stressed that he considers the question of how one should treat disabled infants, who are not yet people, to be separate from the question of how one should treat disabled people, who should, in his view, be given much more societal support to live fulfilling lives. But disability activists critical of Singer point out that the Nazis also started off emphasizing the value of healthy babies. What began with voluntary campaigns promoting birth ended in mass death.

The argument that some lives are inherently, irreparably not worth living does echo Nazi propaganda about people with disabilities, who were described as "life unworthy of life." Beginning in the 1930s, disabled people in Nazi Germany were first prohibited from marrying or having sexual relationships, then institutionalized and compulsorily sterilized, and then murdered in gas chambers. The Nazis did not, of course, stop with the disabled. As McBryde Johnson put it in another essay, the "eugenics dragnet widened . . . concentration strategies, gas-chamber technology and sterilization techniques first designed for disabled people were applied against whole populations defined as genetically undesirable."

This bloody history makes one wary of being too cavalier about the "it"-ness of a fetus, of arguing that some humans are not persons, of preferring some births over others. In the question "Should the baby live?"—and in the statement "I'm glad that baby died"—you can hear a rational consideration of how to maximize well-being and minimize suffering. You can also hear the hiss of hell itself.

The Nazi "Law for the Prevention of Offspring with Hereditary Diseases," which targeted disabled people for compulsory sterilization, was modeled after U.S. legislation, upheld as constitutional in 1924 in the infamous U.S. Supreme Court case *Buck v. Bell*. The young woman at the center of that case, Carrie Buck, had become pregnant after what she described as a rape by her foster family's nephew. Affirming the legality of her sterilization, Justice Oliver Wendell Holmes described the utilitarian calculus underlying his decision: "It is better for all the world if, instead of waiting to execute degenerate offspring for crime or to let them starve for their imbecility, society can prevent those who are manifestly unfit from continuing their kind." He further cited, as precedent, compulsory vaccination laws. Buck was sterilized by tubal ligation in 1927. Ultimately, at least sixty thousand people were involuntarily sterilized, a practice that continued in the United States until the 1980s.

Eugenic sterilization laws were given a veneer of scientific respectability by the work of scientists, who were only barely beginning to understand the mechanisms of heredity, but who were happy to present alleged "evidence" that ethnic, racial, and class differences were due to biological inheritance. The word *eugenics* (meaning "good birth") was originally coined by Francis Galton, a British scientist who developed foundational statistical concepts and was obsessed with rating things on a numerical scale, from the beauty of women to the "eminence" of men. Galton founded a laboratory at University College London that attracted scientists from both sides of the Atlantic, while popular support for eugenics in England reached a point of savage zeal. In 1908, the English novelist D. H. Lawrence fantasized about rounding up people with disabilities and murdering them in a "lethal chamber as big as the Crystal Palace" as a brass band played the Hallelujah Chorus.

Inspired by his meetings with Galton in London, the American biologist Charles Davenport established a Eugenics Research Office in the United States. In 1911, Davenport published an account of their research methods, *The Study of Human Heredity: Methods of Collecting, Charting, and Analyzing Data*, which provided a list of abbreviations for documenting the family histories of those they considered "defective": A for alcoholic, B for blind, C for criminalistic, D for deaf, E for epileptic, F for feeble-minded, G for gonorrheal, I for insane. Davenport's co-authors on this methodological tract were a who's who of American eugenicists. One co-author, the biologist Harry Laughlin, was a Nazi sympathizer who authored legislation for involuntary sterilization—and who corresponded with Margaret Sanger, the founder of Planned Parenthood, who sought alliances with the eugenics movement. She wrote to Laughlin about her goal to "unite the Eugenic Movement and the Birth Control movement, for after all they should be and are the right and left hand of one body."

Another author of *The Study of Human Heredity* was Henry Goddard, an American psychologist who would go on to administer IQ tests to immigrants at Ellis Island. In Goddard's subsequent book, *The Kallikak Family: A Study in the Heredity of*

Feeble-Mindedness, he aimed to show that criminalized and immoral behavior was co-inherited along with poverty and "feeble-mindedness." Goddard's work was later shown to be a chimera of exaggerations and outright lies, and his purported pedigrees were scientifically nonsensical. The alleged "original sin" of the Kallikak family—Martin Kallikak's extramarital dalliance with a lusty barmaid on his way home to his Quaker wife—never happened. The man introduced as Kallikak's illegitimate son was parented by two entirely different people. The part of the family tree supposedly overrun by criminality and indolence actually contained teachers, farmers, and pilots. The few family members who did show problematic behavior appear to have been affected by fetal alcohol syndrome.

But perhaps Goddard's biggest lie—the biggest lie of all the eugenicists—was the idea of "purity." None of us has a bloodline separate from the universal family tree that encompasses all of humanity. None of our families are without sin. In the eugenicists' fables, human biology can always be neatly categorized as "good" or "bad," and bad biology can always be found in a demonized Other. All human fallibility is sequestered in some other sex, race, class, or family, and so can be controlled and eliminated.

Eugenic violence and the projection of inherent badness onto a socially marginalized other continue today. In May 2022, Payton Gendron, an eighteen-year-old white college student, murdered ten Black Americans as they shopped for groceries in Buffalo, New York. Most of his victims were elderly, including Ruth Whitfield, who was survived by seven siblings, four children, nine grandchildren, eight great-grandchildren, and five great-great-grandchildren, a family legacy that would have outraged Gendron. He was obsessed with the idea that whites were being "replaced" through relatively low birth rates, and he intended to "terrorize all nonwhite, non-Christian people and get them to leave the country." Prior to the murders, he penned a 180-page manifesto in which he regurgitated the memes that had radicalized him. Much of the material alleged to show biological differences between races, portraying Black people as a qualitatively distinct, innately dangerous, and inherently inferior "subspecies" of humans. What a grotesque irony, what an evil feat of psychological projection to allege, on the eve of committing mass murder, that *other* people have inherited a predisposition to violence!

One myth that was central to Gendron's racist ideology, and central to the eugenics movement of the twentieth century, was that race has a genetic basis. "Our genetic material is obviously very different," he wrote in his manifesto. This is a false belief, and a dangerous one, but also an alarmingly common one. One study found that middle school students in the United States believed that nearly half of human genetic variation was between racial groups. The reality looks quite different. In fact, nearly all genetic variation exists *within* each racial group, and two people who share the same racial category (e.g., both "Black") may be more genetically different than two people who are seen as being different races. Racial categories are constructed by cultures; they do not reflect categorical genetic differences.

Recent groundbreaking work by Brian Donovan, a biologist and science educator, has found that when high school students are taught accurate beliefs about the genetic similarity of all humans, regardless of race, they show less racial bias. They are, for instance, less likely to agree with statements like "A randomly picked Black person will be unintelligent"—the exact sorts of statements that Gendron listed in his manifesto. We should be careful about attributing murderous hatred to mere ignorance. But I cannot help but wonder: Could Gendron have been inoculated against radicalization if he had received a more sophisticated, more humane genetics education as a young person? Could he and others be turned from hatred in the future?

But that is a moot question for Gendron; the U.S. criminal justice system is largely uninterested in the prospect of moral rehabilitation. At trial, he was found guilty and sentenced to life in prison without the possibility of parole. The judge delivered the sentence as a damnation: "There can be no mercy for you, no understanding, no second chances."

"Many think that eugenics ended with the horrors of the Holocaust. Unfortunately, it did not. The philosophy and the pure evil that motivated Hitler and Nazi Germany to murder millions of innocent lives continues today." Judge Richard Griffin, from the United States Court of Appeals for the Sixth Circuit, wasn't writing about racist hate crimes when he opined that eugenic violence continues today. He was writing about abortion.

The 2021 case before the Sixth Circuit, *Preterm-Cleveland v. McCloud,* focused on an Ohio law that criminalized abortion if the woman's reason for obtaining it was a prenatal diagnosis of Down syndrome. Was such a law allowable given the then–constitutionally protected right to abortion? In the United States, between 60 and 90 percent of pregnancies with fetuses diagnosed with Down syndrome end in abortion (compared to an abortion base rate of 18 percent). In many countries in Europe, rates are even higher, reaching 90 percent in the U.K. and almost 100 percent in Iceland and Denmark, where screening is nearly universal. The Ohio law banned such abortions, and the court allowed the law to take effect. In his concurring opinion, Judge Griffin argued that the modern-day practice of selective abortion was a continuation of the horrors of twentieth-century eugenics.

Antiabortion conservatives were clear that such reason-based bans prohibiting abortion based on disability were a preamble to a total abortion ban. As one conservative legal scholar said, bans that describe a fetus as the object of lethal discrimination "refute the '*it*'-ness of the human fetus.... Whatever makes it intuitively wrong to kill a human fetus because of her race, sex, or disability strongly suggests that it is also wrong to kill the same human fetus for most other reasons." And indeed, the *Dobbs v. Jackson* decision that overturned the legal right to abortion in the United States contained language comparing abortion to eugenics. The state's previous use of force to prevent reproduction was invoked to justify the state's use of force to compel reproduction.

One of the deciding votes in *Dobbs v. Jackson* was Judge Amy Coney Barrett, whose seventh child has Down syndrome. Her vote in *Dobbs* was no surprise. When Barrett was still a judge on the Court of Appeals for the Seventh Circuit, she joined Judge Frank Easterbrook in a dissenting opinion that equated abortion to eugenics—and expanded the conversation about genetic testing. Easterbrook:

> There is a difference between "I don't want a child" and "I want a child, but only a male" or "I want only children whose genes predict success in life."... It is becoming possible to control some aspects of embryos' genomes.... Does the Constitution supply a right to evade regulation by choosing a child's genetic makeup after conception, aborting any fetus whose genes show a likelihood that the child will be short, or nearsighted, or intellectually average, or lack perfect pitch—or be the "wrong" sex or race?

Easterbrook's opinion cited a 2018 *New York Times* article, "Scientists Can Design 'Better' Babies. Should They?," that described how expanding knowledge of the human genome, and expanding access to and use of IVF, have expanded possibilities for prenatal selection. Screening a fetus for Down syndrome is just the tip of the selection iceberg. Now, instead of testing a fetus in utero for a single genetic abnormality that is the definitive cause of a disease or condition, prospective parents can screen up to dozens of in vitro embryos using polygenic scores, statistical summaries of information about thousands or hundreds of thousands of genetic variants that are correlated with the probability of developing a particular characteristic.

As I have described, my colleagues and I have developed a polygenic score that is correlated with the likelihood of developing conduct disorder and ADHD, of becoming addicted to alcohol and opiate drugs, of being convicted of a felony crime. The correlations are small but meaningful; they are about the same size as correlations with lead exposure or child maltreatment.

Ask yourself: Would you screen IVF-created embryos for genes associated with these behaviors to reduce your child's risk? Should you? Should the state be allowed to stop you?

The first human child known to be born from an embryo selected using polygenic scores is named Aurea. Her parents selected the embryo that became her because it had the lowest genetically predicted risk for heart disease. Aurea's parents were already going to use IVF to conceive, because of her mother's age. They were already going to have to choose just one embryo out of several to implant first. (Typically, three to four embryos are created during an IVF cycle, and there can be as many as eight.) And they were already going to screen their embryos genetically: Most IVF-created embryos are screened for Down syndrome and other aneuploidy, an abnormal number of chromosomes. Why stop there? reasoned Aurea's parents. Why leave information on the table?

The first fertility doctor they consulted about using polygenic screening turned them away, saying that it would be unethical. Undeterred, they simply found another IVF clinic that would work with them. One company, Genomic Prediction (motto: "Choice over chance"), now provides genetic screening services to IVF clinics in thirty-seven countries. Another, Orchid ("We help couples have healthy babies"), advertises whole-genome sequencing and tests genetic risk for psychiatric conditions like bipolar disorder and schizophrenia.

Like Aurea's parents, Titan Invictus's parents used IVF to create embryos and used polygenic scores to choose among them. Her mother, Simone Collins, described her reproductive journey in a podcast interview with Steve Hsu, the founder of Genomic Prediction. When she was in her late twenties, she tried to conceive the old-fashioned way and then underwent an escalating series of fertility treatments, but eventually it became clear that without IVF, she would not get pregnant. The Collinses ended up doing five rounds of IVF in a single year, what they called their "harvest year," and ultimately created more than two dozen embryos. As of 2025, the couple has had five babies (they hope to have eight in all), and they used genetic information to pick the children's biological sex, to screen for risk for the cancer that killed Simone Collins's mother, and to get information about each child's risk for schizophrenia. In her interview with Hsu, Collins described herself as an IVF convert: "I love having that control."

Some scientists have called polygenic embryo screening "amoral nonsense" and criticized parents who used it, like the Collinses, as "womb goblins." More dispassionately, the American College of Medical Genetics and Genomics concluded in

May 2023 that "preimplantation genetic testing for disorders that exhibit multigenic or polygenic inheritance is not appropriate for clinical use and should not be offered as direct-to-consumer testing at this time." Among the many scientific and clinical concerns are that genetic predictions can be highly uncertain, that there are large discrepancies in the communities currently represented in genetics research, and that social inequalities in access to expensive reproductive technologies could be inscribed into the genome.

Even as scientists urge caution, companies like Genomic Prediction show that, as a news headline at Bloomberg News proclaimed, "the Pandora's box of embryo testing is officially open." And recent surveys show that the U.S. public is more or less okay with this, at least in theory: A majority (58 percent) say that polygenic embryo selection is either "morally acceptable" or "not a moral issue" when used to test for "medical and nonmedical traits," and 43 percent say they would be willing to use it, if they were already doing IVF, to increase their child's chances of getting into an elite college by a few percentage points.

In contrast, many antiabortion conservatives believe that a fertilized embryo is a person from the moment of conception and is therefore entitled to the same rights as anyone else, including a "right to life." They are fiercely opposed to polygenic embryo selection and, often, to IVF. One article in *The Federalist,* a conservative website, argued that genetically testing embryos "pits unborn siblings against each other, and the highest scorer gets to live." Polygenic scores were described as "technology that replaces the inherent value of life with a numeric placeholder." Similarly, Albert Mohler, president of the Southern Baptist Theological Seminary, warned that polygenic embryo selection, and IVF more generally, compromise human dignity and defy God-given "natural limits upon our creaturely power and authority."

In this way, debates about polygenic screening of embryos are recapitulating debates about abortion. In both, the history of American and Nazi eugenics, which involved the horrific use of state power, is cited as justification for the state's exercise of power over people's efforts to build their families on their own terms.

New technologies surface tensions between values. Here are some of the values I recognize in the debates about prenatal genetic testing: The value of biological diversity. The value of diminishing suffering. The value of considering the whole of a human life, rather than reducing it to a single trait or attribute. The value of a human life in embryonic form. The value of equality. The value of inclusion. The value of sexual pleasure. The value of not being forced to be pregnant. The value of self-determination. The value of giving and receiving care. These values are not in strict opposition but often exist in tension. I mistrust anyone who doesn't hold ambivalence around the bioethical dilemmas posed by our growing knowledge of the human genome.

I was not put into a position to choose between embryos; nature chose for me. If I could have chosen my children, selected them from a menu of physical traits and temperamental quirks and psychological abilities and inclinations, would I have chosen these particular children, with their specific characteristics? Probably not. They are not "easy" children. But I am grateful for qualities I would never have known to wish for. In their variedness and difficulty, they are infinitely more interesting than what I imagined before I was a mother.

In their room, I put a framed print by the artist Hallie Bateman that reads, "It's a miracle we ever met," and it does feel that way to me: Each of them is an astonishment. Out of all the possible people that I could have mothered! And now that we have met, I have the rest of my life to get to know them. Would my wonder and delight at their sheer improbability be diminished if I had deliberately picked each child from a stockpile of embryos, rather than surrendering to the horrifying lack of control that reproduction ordinarily entails? Maybe. But I think it's more likely that any attempt at control on my part would, in the end, only highlight how ultimately laughable the idea of control over creation is. Not even God himself was able to create a child who went according to plan.

At the same time, I *was* able to plan. I was one of the fortunate few in the history of womankind who spent her entire reproductive life with legal contraception and legal abortion. In Texas, at least, this window of history is closing, and it makes me afraid for my daughters. I do not think the state should be able to stop them from using IVF or from making decisions about which embryos to implant using whatever information they deem relevant, even if I find their decisions personally repugnant. I don't think the state should be able to stop their abortions, even if their abortions would break my heart. We ought not force them to grow anything in their garden, even as we ought not force them to pave over their soil.

And still, I recognize the powerful and valid critiques of selective abortion and embryo selection from the disability justice community. It is *dangerous* to assume that some lives are not worth living, to reduce the complexity of human lives to single characteristics. We ought to consider: When we jettison chance in favor of choice, if we try to purge the Garden of wild invasives, what might be destroyed?

Let's consider a terrible thought experiment: A dystopian state attempts to breed the sin genes out of us in its quest to create über-moral man. In every generation, only the most sober, abstemious, scrupulous, rule-bound, pleasure-deferring, careful, conscientious, puritanical among us are allowed to reproduce. All reproduction is accomplished through IVF, so no one must endure the indignities of sexual arousal or the temptations of sexual pleasure. All embryos are screened for genetics found to be correlated with antisocial behavior or any behaviors deemed insufficiently self-controlled, and these genetic predictions are continually tested and revised with every generation.

This type of artificial genetic selection is commonly used in agriculture. In fact, the method for creating polygenic scores is borrowed from agricultural genetics, where it is known by the term *expected breeding values*. If done consistently and at scale for multiple generations, artificial genetic selection can result in massive shifts in the average phenotype. Consider the chicken to be roasted for supper. After decades of selective breeding by industrial chicken farmers, the average chicken in 2005 weighed almost five times more than the average chicken in 1957. What would once be unrecognizably, freakishly extreme has become the new normal.

Chickens have short generational times, which means that effects of selective breeding are seen more quickly than they would be for humans, who reproduce more slowly. And, obviously, agricultural breeders have a degree of absolute control over the reproduction and the rearing environments of animals that we would find ghastly in humans. But the effects of even a few generations of artificial selection in animals show us that, like compound interest, the changes brought about by selection accrue, leading to extreme results that are difficult to imagine at the outset.

So, what if our dystopian state succeeded, through the marvels of genetic science and the horrors of eugenic social control, in breeding the characterological version of the nine-pound roasting chicken, the self-controlled extreme, with none of the clamorous desires of the flesh? The means are horrific, but what about the end? Would it be a good thing for humanity to maximize on self-control? To no longer be so troubled by our insubordinate flesh? To return to a garden of Eden—not the one imagined by queer poets, overrun with wild invasives, but the St. Augustine version?

In Augustine's vision of Eden, humanity was unafflicted by violence, rebellion, greed. There was no theft, no intoxication, no selfishness, no gluttony. No lust, no desire, no risk, no indulgence, no deviance. Adam was not even afflicted with the indignity of involuntary erections. Instead, he could will his cock to harden as some men can deliberately wiggle their ears. A woman could open herself without any sensation at all, except perhaps twinges resembling menstrual cramps. Does this sound appealing? If given the chance to rid humanity of any part of our biology that predisposes us to immorality, would we want to? Or would we discover, as the philosopher Susan Wolf suggested, that "there seems to be a limit to how much morality we can stand"?

The data suggests that we do not prefer St. Augustine's garden. We do not want our sin genes bred out of us. If we did, we wouldn't be having so much sex with people who are a little impulsive, a little risk-taking, a little rule-breaking. I once swiped right on a man's dating profile (that swaggering artist) just because he listed "driving too fast" as one of his interests. He broke my heart, and I wouldn't take it back, and I know I'm not the only one whose sexuality is oriented toward novelty, uncertainty, variety, risk. In our sex lives, we are of the Devil's party. In fact, one of the ways my colleagues and I have gone about discovering genes associated with addiction and antisocial behavior was by studying the genes of people who reported many sexual partners, because the same genes affect all these different manifestations of "bad" behavior. That is, we learned about the biology associated with socially undesirable behavior by studying people who frequently engage in acts of desire. The plant's rebuke to the gardener in Louise Glück's "Clover" could be a rebuke from our genes, sometimes imagined to be unwanted, even as our wants perpetuate their distribution: "You should know / that when you swagger among us / I hear two voices speaking, / one your spirit, one / the acts of your hands."

Those "two voices speaking" can be heard in the genome. *CADM2*, a gene associated with antisocial behavior and addiction, shows evidence of "balancing" selection: Rather than pruning one version of the gene or another out of existence, evolutionary forces have maintained genetic diversity. The ambivalent garden god that is evolution sometimes treats a genetic variant like a nettle, maintaining exceptions that are never permitted to become rules, hoarding a tendril of something it otherwise wants dead. And even if certain genetic variants are universally disadvantageous, the capacity for mutation, for deviation, is the engine of evolution. The garden must contain the seeds of its own destruction for it to self-create.

Émile Durkheim, the French academic who formalized the field of sociology, proposed a paradox similar to that of the garden with regard to human society. He argued that a saintly society without crime would be impossible—and even if it were possible, it would be undesirable. Crime was, despite its appearances, "a factor in public health, an integrative element in any healthy society." A crimeless society, in Durkheim's view, would happen only if everyone were exactly alike in what they value as good and bad and in how they live out their values. Such uniformity was impossible, given that people differed in their social and physical environments and their "hereditary antecedents." Deviation from any norm is inevitable.

But even if such uniformity were possible, Durkheim argued that it would be undesirable, because without diversity, without difference, systems of morality could never progress. A social system that succeeded in stamping out all variety in "good" or "bad" behavior would petrify. A monoculture without the possibility of mutation is an evolutionary dead end; deviation is the grist for evolution. Durkheim called attention to figures in the past who were condemned as criminals when alive, but whose morality we now consider superior to the norms of their age, and who prepared the way for new moralities. Socrates, who was sentenced to die by drinking poison for corrupting the youth of Athens. Jesus, who was sentenced to die by crucifixion for blasphemy and treason. Moral transformations depend on there being space for individual differences in moral behavior to manifest themselves. Durkheim wrote, "But so that the originality of the idealist who dreams of transcending his era may display itself, that of the criminal, which falls short of the age, must also be possible. One does not go without the other." One is the other.

In 1895, Durkheim suggested that the moral entrepreneur could only emerge from a culture that also produced the criminal. More than a century later, Yvon Chouinard, the rock climber turned businessman who founded the company Patagonia, echoed this sentiment, suggesting that the criminal and the entrepreneurial are often one and the same: "If you want to understand the entrepreneur, study the juvenile delinquent. The delinquent is saying with his actions, 'This sucks. I'm going to do my own thing.'" Chouinard's supposition has some empirical support: In the United States, the people most likely to have started a business are those who have benefited from a variety of social advantages (e.g., coming from high-income, two-parent, well-educated families) and who engaged in antisocial behavior as teenagers, such as damaging property, fighting at school, shoplifting, robbery, assault, gambling, and dealing drugs. Crime is costly for society, but the deviance that is reflected in crime, particularly adolescent crime, also generates innovation.

What Durkheim and Chouinard proposed to be true for human societies, cancer biologists have proposed to be true for the multicellular body. In a multicellular body, cells cooperate, benefiting the entire organism, but the body is vulnerable to cheating by individual cells, who hog resources, shirk their tasks, trash their environment, and reproduce without constraint. We call these cheating cells "cancer." The arms race between cancer and the cooperative body is critical for evolution. The possibility of cheating by individual cells puts selective pressure on the body to innovate new strategies for detecting and neutralizing cheaters, and sometimes what initially looks like cheating can become the new normal.

All cooperative schemes, then, require an element of what would, in excess, cause their dissolution. An individual deviating from the norm risks harming the collective. But without deviation, without mutation, there is no innovation, no evolution. The eugenic fantasy that "bad" biology can be clearly identified, cleanly sequestered, coercively eliminated, is just that—a fantasy. Varieties of life do not conform to our neat categories of good and bad. The solution to the problem of suffering—to evil—will not be finally found by tinkering with the forces of reproduction. There is no solution. We can only imagine further evolution, new innovations for cooperation and repair.

I've skipped ahead again, from the end of my first marriage to the child of my second. The birth of a child can feel redemptive: How could actions and experiences that resulted in this precious good be entirely bad? But what allowed me to begin to forgive myself for the end of my religious belief, and the end of my first marriage, and the disappointment that both events brought my parents, was not a new baby. It was a new puppy, who reminded me of what we owe our fellow creatures.

Consequences

In which I adopt a fearless puppy and discover that the gentle methods of his socialization contrast starkly with the cruel, futile punishments imposed on children and prisoners

IT IS THE SUMMER of 2019. I am in the middle of moving houses, from the tiny rental house I was sharing with my two children during my divorce from their father to the somewhat less tiny house I have purchased, a duplex on the poorest street of a rich neighborhood. My ex-husband is keeping the home we had shared. His mother has written me a six-figure check—more than I have ever seen in one place, but less than half the value of the house—to buy me out. The money is enough for a down payment on a new house near the children's school, but not enough to buy furniture. The new house is empty, except for our beds and our books and my piano.

Taking the children to ballet class the day before the movers arrive, we wander into the farmers market. My thought was that flowers would make the empty house feel more cheerful. Instead, I find myself at a booth run by a local animal shelter. A litter of puppies has been found on the side of Highway 71. One catches my eye. He is different from his siblings, black-and-tan rather than their black-and-white. As I chat with the woman running the booth, I take my eyes off my children, and they climb into the puppies' enclosure. They, too, are drawn to the outlier. I find myself adopting him—signing paperwork, handing over my credit card, aware on some level that adopting a puppy the day before moving is ludicrous, aware on another level that I don't care. The lack of furniture suddenly seems like an asset: There's very little for a new puppy to chew. The prospect of an empty house suddenly seems less dire.

(He did, of course, find things to chew—baseboards, cupboards, door frames, pantry shelves.)

It takes me more time than it should to realize what is obvious to most people immediately: Guinness will not remain a small dog. He gains ten pounds in the first month we own him. He pulls at the leash, jumps on visitors, growls at strangers passing by the house, and constantly patrols the perimeter of the house, ceaselessly pacing from window to window and door to door. Later, I genotype him and learn he is a mix of Great Pyrenees, Chow Chow, Rottweiler, German Shepherd, and Shar Pei. He is, in other words, a hybrid of the world's guard dogs, bred for centuries to protect property and livestock.

I look up every breed in his lineage and the advice is always the same: Early training and socialization are essential. Dogs of these breeds tend to be large, powerful, independent, intelligent, loyal to their owners, fearless, vigilant to potential threat, and aloof or aggressive with strangers. Guinness jumps over a porch railing and off a second-story roof in his fury at another dog being on the front lawn. (No one, thankfully, is injured.) I realize, much too late, I need help. I hire a dog trainer.

In our first session, the trainer outfits Guinness with an electronic collar and hands me its remote control. We are working on basic commands: sit and stay. I am instructed to push the button that activates Guinness's electronic collar every time I say a focal command word. The sound of my voice saying "Sit" is paired with a mild electric stimulation to his neck. The trainer lets me feel the stimulation on my own palm to reassure me that we are not causing Guinness any pain. As promised, the sensation is not painful, just annoying. The collar's buzz is unignorable, like the insistent beep of your car when your seatbelt is unbuckled.

This is classical conditioning. My voice command (the "conditioned stimulus") is paired with an intrusive physical sensation to which Guinness cannot help but pay attention (the "unconditioned stimulus"). The goal is for him to orient his attention to me even when I am not using the collar; we want his attention to be automatic. I've never had a dog before, but I have had, at this point, two toddlers, so the process is not entirely unfamiliar. Children's socialization, as with dogs, depends on classical conditioning for its development.

If you have spent time with a toddler, you already know that children do not come into this world observing moral rules. They have the most bestial of weapons—teeth and claws—and they will happily wield them against another child. Physical aggression is such a common problem among young children that one of the bestselling board books for tots has the ironically inaccurate title *Teeth Are Not for Biting*. It includes lines like "When you feel mad or sad or cranky, you may want to bite / But OUCH! Biting hurts." The bite victims, several of them adults, are drawn with exaggerated tears and mournful expressions. Their "OUCH!" is the biggest font on the page. The book directs children's attention, again and again, to a rule about what not to do (the command) and pairs it with a depiction of another person's pain.

Within days of being born, most infants already show some automatic response at another person's distress, a rudimentary form of emotional empathy. Crying is contagious in nurseries, and not just because babies don't like loud sounds. They won't cry in the same way in response to a chimpanzee's cry—only to the cries of other babies. Human unhappiness seems to be, for most children, innately aversive. Most of us are born with an internal version of an electronic collar, which buzzes and gets our attention when another human is in pain.

As children get older, their emotional empathy for other people's distress is evident throughout their bodies. The physiological stress response begins and their pupils get wider, their hearts beat faster, their skin conducts more electricity as they start to sweat a bit. This early emotional empathy seems to precede the development of cognitive empathy, the ability to name other people's emotions and their likely causes. ("That person is sad because their mommy is gone.") We can intuit that someone is upset long before we can name or categorize the nature of that distress. By adulthood, seeing another person suffer activates multiple areas of the brain, including the anterior insula, which is engaged when experiencing pain oneself.

For most children, then, some level of emotional empathy is "unconditioned"—it doesn't have to be directly taught. And just as I use the automatically attention-grabbing effects of an electronic collar to teach my dog to pay attention to my commands, adults use children's unconditioned distress at another's pain to teach them to pay attention to moral rules. The most common thing that caregivers of young children will do, when the child has transgressed, is associate a behavioral demand with another person's pain. (Ouch! Biting hurts!) As deliberately as we teach dogs to pay attention to our command to "Sit," we teach children to pay attention to our commandments: "Thou shalt not bite."

But not all children experience affective empathy for others' distress to the same degree. Even among children with typical development, there are differences. Children who are more temperamentally fearless, who are less bothered by novelty or intensity or strangers, whose brains seem less reactive to potential threats, show less affective empathy than more fearful children. And at the extreme low end, some children seem almost entirely lacking in emotional empathy. They seem cold, indifferent, detached. Their parents might point out that their biting has caused another child to cry, but the emotional buzz that ordinarily compels attention to that rule and makes it personally meaningful is absent.

Without this emotional buzz, socialization is more difficult. It is, of course, possible to train a dog without an electronic collar, but it is slower, more difficult, more effortful. It requires more consistency and skill on the part of the one doing the training. In the same way, moral socialization is more difficult with children who are low in emotional empathy. They are asked to learn and abide by abstract moral rules. As psychologist Robert Hare described, "It is the emotionally charged thoughts, images, and internal dialogue that give the 'bite' to conscience, account for its powerful control over behavior." Without emotional resonance, "conscience is little more than an intellectual awareness of rules others make up—empty words." Empty words are not always enough: Lower levels of cognitive and affective empathy are unique risk factors for engaging in violent and antisocial acts in adolescence and adulthood.

Once Guinness has begun to learn that my voice commands are associated with something that captures his attention, we begin giving rewards for desired behaviors. I say "Sit" and activate his collar, and then when he sits, I give him a little peanut butter treat and enthusiastic praise: "Good boy! Good boy!" Guinness is strongly motivated by food, even more by praise. "An animal that likes approval this much can be taught to do anything," the trainer says cheerfully. (I flinch in recognition.)

This part of training is also straight out of my Intro Psych class. This process is called instrumental conditioning. We are aiming to increase or decrease the frequency of certain behaviors by giving rewards. We are putting into practice what the early-twentieth-century psychologist Edward Thorndike called the "Law of Effect." If an animal does something closely followed by "satisfaction," that behavior will be more likely to recur, whereas behaviors closely followed by "discomfort" will be less likely to recur. What gets rewarded gets repeated. What gets punished, the idea goes, gets deterred. Consequences matter.

The Law of Effect depends on ancient biological circuits. Rewards, particularly new or unexpected ones, trigger the release of dopamine by neurons originating deep inside the brain. The animal feels pleasure and learns that the action, and the environment in which it is performed, is something to be repeated. These circuits are necessary for animal life. Our brains need signals that are difficult to ignore, sometimes impossible to forget, to motivate the behaviors necessary for survival and reproduction. Eating sugar feels good. So does nursing your baby, cuddling your mother, being admired and accepted by your peers, having sex. Without learning from these signals of pleasure, we might wander about, forgetting to eat or drink; kissing would be like flossing, just another tedious habit.

Just as critical is the ability to learn from pain, which experts define as "unpleasant sensory and emotional experience associated with, or resembling that associated with, actual or potential tissue damage." Experts debate which animals actually experience pain, per se, which is a subjective experience that implies a conscious mental life. However, even creatures that have little to no conscious awareness and don't have the elaborate brain structures that produce the human experience of pain have nociception, the ability to sense injury or potential injury, sensations that lead to organized defensive responses and to behaviors that avoid future injury. Consider the lobster, being boiled for supper, pathetically clinging to the edge of the pot. Its nervous system is primitive, but it still senses injury and works to avoid it.

Nociception is necessary for punishment learning; punishment learning is necessary for life. Fruit fly larvae who lack nociception do not twist and turn to escape from parasitic wasps that inject their eggs into the flies' bodies. The wasps, when hatched, then eat the larvae alive from the inside out. Children who have mutations in their *SCN9A* gene or in their *NTRK1* gene have congenital insensitivity to pain. They play with glass and fire. They mutilate their own fingers. They typically don't live past the age of twenty-five, because they cannot learn to avoid injury. The short lives of organisms who are totally insensitive to punishment testify to the truth spoken by the Chorus in Aeschylus's *Agamemnon:* "For Zeus' law is first in all the world / The law is this: No wisdom without pain."

There is no learning without the ability to feel pain, but pain alone is not sufficient for learning. One of the most famous psychologists of the twentieth century, B. F. Skinner, spent years carefully observing how pigeons and other animals responded to rewards and punishments, food and shock. When he first started conducting experiments on the Law of Effect, he "had assumed, along with almost everyone else, that punishment was simply the opposite of reward. You rewarded people to make them more likely and you punished them to make them less likely to behave in a given way." But by the mid-1930s, Skinner's data had made him suspect that reality was more complicated. Pain and pleasure no longer seemed like strictly opposing forces.

By the 1940s, after a series of studies on the effects of electric shock in rats and pigeons, Skinner abandoned the idea that punishment was the symmetrical opponent of reward. If punishments are immediately and consistently applied after a behavior, the behavior might stop, at least for a time: A man who puts his hand to a hot stove will stop touching it. A child pinched for laughing in church will stop laughing. But often behavior rebounds as soon as the threat of punishment is lifted: A man can just as quickly learn that the stove is cold again. And sometimes punishments lead to no apparent behavior change. A rat who has learned that pushing a lever will give him alcohol might keep on pushing it even when the lever no longer delivers alcohol and now delivers electric shock. This indifference to punishment is especially likely if the punishment is inconsistent or delayed, or if the punished behavior is very habitual, or if the animal doesn't have any other behavioral options, or if the animal is frightened. Even more curiously, sometimes animals show paradoxical responses to punishment: The punished behavior becomes *more* frequent.

In 1948, Skinner published a novel, *Walden Two*, about a utopian community run by a former academic using the principles of classical and instrumental conditioning. A society built on the science of behavior would not have prisons, he emphasized, not because people could be turned into saints but because the crude "threat of pain" doesn't effectively stop them from being sinners. As he wrote: "The old school made the amazing mistake of supposing that . . . by removing a situation a person likes or setting up one he doesn't like—in other words punishing him—

it was possible to reduce the probability that he would behave in a given way again. That simply doesn't hold."

What Skinner concluded from his experiments with small animals, many parents have learned from their experiences with their children: Punishment is less effective than reward. In one study, the psychologist Luke Hyde and his colleagues studied adopted children when they were just over two years old. They had all been adopted shortly after birth, half of them within just two days, meaning that their biological mothers had not raised them or, in many cases, even interacted with them. Yet their biological mothers' behavior predicted the young children's moral emotions. Mothers who had engaged in antisocial behavior, such as breaking and entering or using a weapon, had biological children who, on average, showed less affection toward family members, less fear, and less guilt or distress after they had hurt someone else. Their adopted mothers also rated them as slow to learn from punishment. They would continue to do something even if they had experienced negative consequences for it.

The relationship between biological mothers' antisocial behavior and the callous-unemotionality of their adopted-away children is evidence that something is being passed down from mother to child. Crucially, however, the effect of that inheritance depended on the parenting of the adoptive mother—and not on her use of punishment but on her warmth. Children whose adoptive mothers rewarded them with praise, approval, and affection for behaving in cooperative and helpful ways showed more prosocial emotions, and the effect of this "positive reinforcement" was biggest for children at highest biological risk. Other studies have affirmed the importance of positive parenting for children: Teaching parents, for instance, to use rewards rather than punishments, and to coach and praise their children when they pay attention to other people's emotions, is currently one of the most effective treatments to decrease conduct problems in children.

Reading these studies, I am struck by the simplicity of so many of the examples of positive parenting. "Thank you so much for helping me!" "I love working with you." "Good job!" It makes a difference for children to hear the encouraging statements that a dog trainer will say to puppies. Dogs are not the only animals hungry for approval, for connection, for love.

I doubt my dog's trainer has read the psychological literature on parenting for children with low empathy and low fear, but his advice is remarkably similar. In our training classes, we are coached on how to maintain clear boundaries, how to establish consistent and immediate consequences. Desired behaviors get enthusiastic praise and treats; unwanted behaviors get swift redirection. At the same time, he repeatedly emphasizes that arbitrary or harsh punishment for unwanted behaviors would be not only cruel and inhumane but also ineffective or even counterproductive. We are never to hit or kick our dogs, never to shout at them or spray them with cold water, never to starve them or cage them for too long. We are, instead, to focus primarily on rewarding the behavior that we want to see. Consequences for unwanted behavior must be immediate and must be a message that the dog can learn from. (The children and I, for example, practice saying "no" in a firm voice, turning our backs and averting our eyes when he nips at them; he quickly stops.)

Above all, we must remember that dogs are social animals attuned to relationship, attentive to status and dominance. Affection, praise, presence, and play build trust and connection. A dog who will jump off a roof to respond to a perceived threat feels little fear, is undeterred by pain. The threat of yet more pain, or attempts to instill more fear, will be futile. He will stay instead of jump because of something stronger than fear—attachment.

In 2023, a dogfighting ring was broken up in Bastrop, Texas, near where Guinness and his siblings were found. Twenty dogs were recovered. They were bloody, scarred, chained. Vicious and terrified. I wonder about what would have become of my dog if he had not been rescued, if he had been raised to fight. I shudder to imagine his best features—his strength, size, loyalty, pride, vigilance, courage—warped by cruelty.

As I said, the principles of dog training are the stuff of Introductory Psychology, a class that I've taught more than a dozen times. So why did learning to train a dog feel revelatory? Because every week the trainer emphasized that hitting and screaming at this creature would be cruel and ineffective, and I was once a child who was hit and screamed at. But here, a new sort of evidence: *You shouldn't even treat a dog like that.* Another axiom proved upon the pulses of an animal.

My father always used a leather belt, and the whole ordeal was prefaced by what he considered an appropriate amount of ceremony. Upon returning from a work trip, he would get an accounting from my mother of our misdeeds. We were ushered into my parents' bedroom, where he would open his Bible, the one with his name embossed on the cover, and read to us, usually from Proverbs. "Foolishness is bound up in the heart of a child; the rod of correction will drive it far from him."

And then into my parents' bathroom, where my dad kept his belts hanging neatly inside the "his" of their "his and hers" closets. Despite the rigamarole, his heart was never in it. He would stop after just a few lashes, especially if we cried. I have no pride when it comes to pain, or even the threat of it, and would start hollering before the first blow landed.

My father is a tall man with a military bearing and a stoic demeanor; you would think he would be the one we feared. And we did, to a point. But we knew our mother was the one to watch out for. She never used a belt. She preferred a wooden spoon or, worse, one of her sandals, with their smooth, flexible, stinging soles. No ceremony was necessary: She could whip her shoe off her tiny foot faster than we could apologize or flee. She was the one, I felt as a child, who wanted to hear us cry.

This is an ordinary memory. In 1993, when I was eleven years old, 50 percent of U.S. parents with a child between the ages of two and twelve reported spanking their child. Since then, that number has dropped. By 2017, only 35 percent of U.S. parents said they spanked their children, and a 2021 study in the U.K. found that only 20 percent of ten-year-olds had experienced physical punishment. "Only"—that's still millions upon millions of children. And many adults approve of corporal punishment in theory, even if they don't practice it themselves; 74 percent of millennials in the United States agree with the statement that "a good hard spanking is sometimes necessary." In contrast, Scotland and Wales have made spanking illegal, and more than 71 percent of English adults say that smacking, hitting, slapping, or shaking a child is unacceptable. The United Nations has called the "clear and unconditional prohibition of all corporal punishment" a "necessary" legal reform for reducing violence in a society, but the United States remains the only country in the world that has not ratified the United Nations Convention on the Rights of the Child.

The effect of spanking on kids is one of the most thoroughly investigated topics in all of child psychology, and the evidence is clear that spanking does nothing but hurt children. One paper surveyed the results of all previous research on spanking, pulling together data on 160,927 children. There were seventy-nine "statistically significant" correlations between spanking and child outcomes. Of these, seventy-eight were negative, meaning that spanked children showed worse outcomes. They had worse relationships with their parents, worse anxiety and depression in childhood, greater aggression and antisocial behavior in adolescence and adulthood. For many outcomes, spanking was about as strongly associated with bad child outcomes as physical abuse is.

These results can also be seen in children's brains. In one study, eleven-year-old children were put in a brain scanner and shown pictures of adults whose facial expressions displayed subtle signs of emotion: fear, happiness, sadness, anger. Compared to children who had not been spanked, children whose parents used corporal punishment showed greater activation in multiple areas of the prefrontal cortex in response to subtle emotional cues, the same sort of hyper-responsiveness shown by children who had been physically abused. Children who have something to fear from their parents get very good at watching adults' faces, scanning for signs of a fast-gathering storm, for signs that they need to take cover.

Ironically, spanked children are no different than children who are not spanked in one way: They are not more compliant with parental demands. Spanking, for all its potential harms, doesn't accomplish its ostensible purpose of making children obedient. This conclusion would be no surprise to B. F. Skinner, but it has not penetrated American consciousness. In every demographic group, the majority view is that spanking is an acceptable, and sometimes necessary, form of punishment for children. Support for spanking is higher among people who live in the South than in other regions; among Republicans than Democrats; among African Americans than other races and ethnicities; and, most of all, among people who identify as "born-again" Christians.

Enthusiasm among U.S. Christians for hitting their kids has a long history. Puritan colonists thought that inflicting pain was necessary for combatting children's inherent wickedness. Not even infants were spared, because not even infants were innocent. Original sin demanded the mortification of the body. Esther Edwards Burr, the daughter of Jonathan Edwards, who championed the idea of universal depravity in his 1758 polemic *The Great Christian Doctrine of Original Sin Defended*, once wrote her parents to give a pleased account of how quickly her nine-month-old responded to whipping. (Corporal punishment was not sufficient, even in the Edwards family, to guarantee that children would avoid antisocial violence: Esther's son, Aaron Burr, became an infamous murderer.)

Today, Christians can choose among an array of books describing methods for corporal punishment: *Shepherding a Child's Heart; What the Bible Says About Child Training; Spanking: A Loving Discipline; The New Strong-Willed Child;* and— who could forget this title—*God, the Rod, and Your Child's Bod*. The books vary in their recommended implements (plastic tubing, wooden spoons, paddles) and timelines (some beginning at six months old, others waiting until fifteen months). But all justify hitting children as a necessary response to the child's inherent—and inherited—wickedness. Ginger Plowman, for instance, in her 2003 book *Don't Make Me Count to Three: A Mom's Look at Heart-Oriented Discipline*, wrote: "To blame sin on a child's heritage is to state the obvious. In one sense, all sin is hereditary. We inherited it from Adam. But rest assured, the biblical use of the rod helps to deal with any hereditary trait that needs correcting." Similarly, after his 1994 book *To Train Up a Child* was circumstantially linked to the deaths of three children, all from Christian homeschool families, the Memphis-based pastor Michael Pearl said that he would never give up support for corporal punishment: "To give up the rod is to give up our views of human nature, God, and eternity."

I do not hit my children, but I am often overwhelmed by them, by their fearlessness, by their sheer loudness. I must work to avoid yelling in desperate frustration and am sometimes momentarily bewildered by what parenting books mean by "accountability" and "boundaries" and "consequences" in concrete terms, if they don't mean hitting and being hit. I don't think it's a coincidence that the title of one of the most popular parenting books in the United States, especially among millennial parents who want to relate to their children differently than their parents related to them, is *Good Inside*—the exact opposite premise of the doctrine of original sin. Without the assumption of children's inherent depravity, spanking and other forms of harsh punishment lose their rationale. But those of us who were raised with harsh punishment lack an internal working model of what healthy boundaries look like. We need new skills.

One last result about the effects of corporal punishment: Children who were spanked are more likely to be the victims of domestic abuse as adults. One interpretation of this result is that the experience of corporal punishment "wires" attachment and affection together with violence and fear. If you were spanked, it might feel normal when someone who hugs and kisses and cuddles you, who claims to love you, also hurts you and tells you that hurting you is love. Spanking by the person on whom you are most dependent can train you to empathize with the aggressor and scorn any sign of dependence in yourself or others. In the novel *Anthem*, the writer Noah Hawley described how we recreate our childhood hurts in our adult relationships. How we run toward pain.

You act like the world is filled with rational choices, that our brains are simple binary systems where if you push the red button and get a shock, you stop pressing the red button. But what if the red button pushes you? What if your father is a red button and he trains you to push him, rewards you for pushing him? What if he teaches you that pushing the red button is called love, a sharp electric shock that leaves behind a dull ache? Doesn't it make sense that when you finally escape into the wild, and you see a red button, you would push it, because who doesn't want love?

If you weren't raised by parents who hit you and told you it was love, this might all sound unfortunate but not personally relevant. But the ideas that undergird the widespread practice of corporal punishment in Christian families are the same ideas that helped build mass incarceration in America. Millions of people live, both metaphorically and literally, in the edifices constructed by people convinced that the only recourse for people's inherited fallibilities is ever harsher punishment.

In 1970, James Dobson, an Evangelical psychologist and author, published *Dare to Discipline*, which called for a renewed commitment to spanking in Christian homes. And interspersed with advice about spanking, bed-wetting, sex education, and chore charts, Dobson also talked about crime, criticizing lenient judges: "Lawlessness is more likely to occur in a society where crime appears to pay." Christians had long been exhorted to use physical discipline, but now spanking was given an updated rationale, one that explicitly tied its use in the home to support for more punitive criminal justice policies.

Dare to Discipline sold millions of copies, even earning a spot in the Nixon White House library, and launched Dobson's career as author, radio personality, and conservative organizer. The book's popularity not only reflected the growing numbers of Americans who identified as "born-again" or Evangelical Christians but also the centrality of punishment to white Evangelicals' ascendent political influence. In *God's Law and Order: The Politics of Punishment in Evangelical America,* a "religious history of mass incarceration," historian Aaron Griffith describes how white Evangelical leaders organized around punishment: "Crime and punishment . . . were central to [Evangelical] entry into American public life. . . . Evangelicals not only lobbied for policies and voted for politicians who helped build America's carceral state, they also helped make these changes appealing to other citizens." *Dare to Discipline* illustrated one way that harsh criminal punishments could be made appealing to millions of American citizens: by analogizing the carceral state to the good parent, who hits his child because he loves him.

In 1970, when *Dare to Discipline* was published, the incarceration rate in the United States was 161 per 100,000 people. In 1971, President Nixon declared the War on Drugs, and then President Reagan escalated it. Republicans and Democrats competed to see who could punish more. More money was poured into more police, more militarization, more surveillance, more prisons; more acts were defined as crimes; more crimes carried sentences of more years in prison, and more crimes could be punished by the death penalty. The incarceration rate in the United States finally peaked in 2008, at 755 per 100,000—the highest rate in the world by far. For comparison, the U.K. in 2022 incarcerated people at a rate lower than 1970s America.

In addition to being supported by Evangelical Christians, the policies of mass incarceration were further legitimized by academic social scientists, who began to apply the usual tool kit of economics—the study of wages and prices—to understanding crime. In 1968, twenty years after psychologists had documented in animal studies that behavior did not reliably decrease in response to punishment, that punishment often backfired, the economist Gary Becker, who would go on to win a Nobel Prize for his application of economic methods to human behavior, proposed in a paper entitled "Crime and Punishment: An Economic Approach" that punishing people more harshly would decrease crime. He thought he knew how humans would respond to the threat of longer prison sentences, because he knew how humans responded to higher prices. He presumed that the world is filled with rational choices, that our brains are simple binary systems where if you push the red button and get a shock, you stop pressing the red button. He predicted that, if the "price" of crime went up—if, for instance, a person convicted of a minor drug crime was kept in prison for five years instead of two—crime would inevitably go down.

The economic analysis of crime and punishment that Becker pioneered became extraordinarily influential, not just in the academy but in the halls of power. His research assistant on the paper, Isaac Ehrlich, subsequently applied economic methods to the study of the death penalty, arguing in the *American Economic Review* in 1975 that the death penalty deterred murder. The next year, the U.S. Supreme Court ruled in *Gregg v. Georgia* that the death penalty was not "cruel and unusual punishment," allowing capital punishment to continue in the United States. (The death penalty is outlawed in the U.K., and no one has been executed there for over sixty years.) In his opinion, Justice Potter Stewart, writing for the plurality, cited Ehrlich's paper, among others, in support of the conclusion that "the infliction of death . . . is not without justification."

Another leading academic voice advocating for more prisons and harsher punishments in this era was James Q. Wilson, a political scientist who became known for his "broken windows" theory of policing. Alarmed by how crime rates had increased throughout the 1960s, Wilson argued that the liberal focus on the "root causes" of crime, particularly poverty, was muddle-headed and ineffective. The key to reducing crime was punishing more. In *Thinking About Crime*, published in 1975, Wilson described "tough-on-crime" policies as simple common sense—just like "smacking" your children.

In the 1980s, Wilson further expanded his theory of crime and punishment in *Crime and Human Nature*, co-authored with Richard Herrnstein, a psychologist who had done his PhD on pigeon reward schedules under the supervision of B. F. Skinner himself. *Crime and Human Nature* discussed "constitutional factors"—individual differences in genetics, personality, and physiology—that might predispose some people to crime. But just as in Christian parenting books, constitutional factors never mitigated how severely an individual should be punished. Instead, they made severe punishment even more necessary. If some people are especially likely to be fallible, they argued, rehabilitation programs wouldn't work. The only solution was incapacitation via longer prison sentences. In Wilson's words: "Wicked people exist, and nothing avails except to set them apart from innocent people."

Central to Wilson's and Herrnstein's (and Becker's) model of crime was the concept of "choice," where choice was (re)defined as behavior that responded to potential incentives, whether those incentives were higher prices or longer prison sentences. "When we say people 'choose,' we do not necessarily mean that they consciously deliberate about what to do," Wilson and Herrnstein wrote. "All we mean is that their behavior is determined by its consequences."

To Skinner, the fact that people, like other animals, respond in predictable ways to rewards and punishments was further evidence that "choice" was an illusion. You could not really have done otherwise than you did, in his view, because your behavior was determined by a phenomenal web of cause and effect. This view of choice had implications for punishment: Immoral or criminal behavior was best understood as a failure of learning, not as a depravity of will. If a child continues to hit his sister even if he is spanked, then the incentives are faulty, not the child.

In contrast, Wilson and Herrnstein did not interpret the fact that people respond to rewards, and sometimes to punishments, as evidence that choice is an illusion. This was, instead, the very *definition* of "choice": Choice is behavior that is determined by its consequences. If a dog came when he was called because you previously rewarded him for coming, then the dog, under this definition of choice, "chose" to come. But even as they adopted a definition of "choice" that did not at all involve freedom from the push and pull of factors beyond a person's control, Wilson and Herrnstein retained the retributive idea that people deserve to be punished for their choices.

Choice was defined in terms of people's response to rewards and punishments; rewards and punishments were justified in terms of people's choices. This circularity allowed punishment to become its own justification, a closed loop where the answer is always to punish more. The threat of harsh punishment was presumed necessary to prevent people from behaving badly, but when the threat of harsh punishment did not, in fact, prevent people from behaving badly, people were seen as deserving to be punished even more harshly. This was the twisted logic of mass incarceration, a secularized version of the logic of hell. And it was rooted in a view of human nature that is strikingly Augustinian, although it was of course described in entirely secular terms: What you inherit might corrupt the quality of your will, bend it toward vice, but the fact that this inheritance is beyond your control does not in the least diminish how severely you should be punished; it just makes your suffering even more necessary.

These ideas were not merely academic thought exercises: As Wilson and Herrnstein were writing the book, which they began in 1977, Wilson served on the Attorney General's Task Force on Violent Crime under President Reagan, which recommended the abolition of parole and bail for certain prisoners, mandatory minimum sentences for some crimes, and, above all, more prisons. These were the policies that ultimately caged more people than had ever been incarcerated in the history of the world.

Wilson's ideas—reverberations of the doctrine of original sin—can also be heard in Justice Antonin Scalia's line of questioning in the 2011 U.S. Supreme Court case *Miller v. Alabama,* which focused on whether adolescents convicted of homicide could be given a mandatory sentence of life in prison without the possibility of parole. Responding to Bryan Stevenson, the lawyer representing two boys who committed murders when they were fourteen years old, Scalia asserted: "Well, I thought that modern penology [the academic study of crime and punishment] has abandoned that rehabilitation thing." In other words, even adolescents have no hope of transformation. The United States remains the only country in the world that allows youth to be sentenced to life without parole.

In 2015, the libertarian economist Alex Tabarrok summarized the legacy of the more-punishment-less-crime model: "We have now tried that experiment and it didn't work." Locking up more people does not lead to decreases in crime. While incarceration rates in the United States quadrupled between 1960 and 1990 and doubled in the U.K., they fell by more than half in Finland and remained stable in Germany—but the crime rates in these countries during that time followed nearly identical patterns. As Michelle Alexander described in *The New Jim Crow: Mass Incarceration in the Age of Colorblindness,* "tough on crime" policies built a staggeringly unequal system of social control: The incarceration rate among Black Americans is higher than the rate of Black incarceration in South Africa at the height of apartheid, and higher than the total rate of incarceration in the Soviet Union at the height of the Gulag. "We are gradually discovering—at an untold cost in human suffering—that in the long run punishment doesn't reduce the probability that an act will occur." Skinner came to that conclusion in 1948, and we still have not learned the lesson.

As I write this chapter, a close friend texts me. There has been a brutal murder in Memphis. He feels helpless, scared, outraged. He has ideas about how the city could be made safer, all of which involve harsher criminal punishment. If people were "treated as terrorists," if more people received the death penalty, "that would put a stop to this revolving door of criminals."

I recognize in his frustration an idea that comes readily to me, too—that consequences must be cruel to be effective. Like him, I was raised to believe that behavior was always a reflection of an innately wicked nature, that fragile virtue was best secured by remembering that a belt awaited me in this life and a lake of fire in the next. Steeped in the doctrine of original sin, it can be difficult to remember that most countries have laws and enforce them, have policies to encourage virtue and discourage vice, without pursuing ever harsher punishments for ever expanding numbers of people, as the United States does. But I also know that the threat of draconian punishment does not change people for the better, any more than spanking a child or hitting a cornered dog does. There is no wisdom in this life without pain, but inflicting more pain is rarely wise. I respond:

> people have been saying that if they are just more punitive and bring back the death penalty then that would nip this crime thing in the bud for literally 60 years
> and it does not work, it has not worked, it will not work

He is unconvinced:

> so we should just let them keep committing crimes?

I interpret his question as a plea: *If harsh punishment isn't the answer, help me understand what is.*

Mark Kleiman was a public policy expert who was strongly influenced, in his early career in the 1980s and 1990s, by Becker's and Wilson's analysis of crime and punishment. He spent years arguing that more punishment, and thus more prisons, were necessary to have less crime. Only later, as crime went down but incarceration kept going up, and up, and up, and up some more, did he come to realize how punitiveness can fall short. In his book *When Brute Force Fails,* he wrote about which policies might actually deliver on the promise to reduce crime, including preventative measures like lead abatement, high-quality preschools, nurse visits for expectant mothers, and access to psychiatric medication. Other social scientists, like my friend Jennifer Doleac, have identified numerous "forks in the road" where effective accountability actually means less punitiveness: dropping the charges for first-time nonviolent offenders, expanding access to prison education, diverting juveniles to restorative justice programs, subsidizing wages to make low-income jobs pay more.

Similarly, philosopher Hanna Pickard, drawing on her clinical training with psychiatric patients struggling with addictive and personality disorders, has described how effective treatments for people who harm themselves and others involve what she calls "responsibility without blame." Demanding that someone reflect on the reasons they behaved as they did, calling their attention to the natural unwanted consequences of their behavior, imposing agreed-upon consequences to encourage motivation, deliberately practicing alternative ways of thinking and behaving, and setting clear boundaries and expectations about how behaviors will be responded to in the future, while also maintaining a stance of empathy and concern, are core components of therapies that can produce transformational change. In this way, effective psychotherapy, like good parenting, is neither permissive nor ruthless.

But designing a system for true accountability, rather than just punishment for punishment's sake, brings us back to the question of choice—the choice between consequences that are effective versus those that satisfy our thirst for retribution. As Kleiman wrote: "The first step in getting away from brute force is to *want* to get away from brute force: to care more about reducing crime than about punishing criminals, and to be willing to choose safety over vengeance when the two are in tension."

The word *want* here is apt. While the apparatus for mass incarceration in the United States was intellectually justified in rational terms, using the mathematics of incentives and the rhetoric of deterrence, it was unconsciously motivated, I believe, by the same reasons that some parents want to hit their children—not because it "works" to reduce an unwanted behavior but because it satisfies our feelings of rage and powerlessness, and because we ourselves were trained to empathize with the aggressor. We want vengeance. And as we will see, even though punishment is often justified as a way to limit our animal instincts, the lust for punishment is yet another bestial desire, as rooted in our evolutionary history as the sins we punish.

Retribution

In which I, reflecting on the harm we inflict on those closest to us, examine the evolutionary origins of resentment and blame—and the creatureliness of forgiveness

LET ME TELL YOU another fable.

Once upon a time, a group of sisters were banished from their home. They had all been raised together in the old castle, their childhood home. But the old castle had been destroyed. Their mother, the old queen, was dead, and they had all been exiled from their homeland.

After they were banished, the sisters fought and fought until they nearly destroyed each other. The sister who came to be called the new queen was not the tallest, nor the strongest. But she had the dark beauty of their mother, the dead queen, and she never let anyone see her wounded or vulnerable. Things were calmer once they all accepted the inevitable, that she in her loveliness would be the new queen. Banding together at last, they survived the long winter and prepared to start again.

The sisters built their castle with many chambers and tunnels, and they attacked any stranger that came near. They took suitors and began to have children. The castle grew larger every day.

But then the queen issued a terrible edict. Her sisters would pay her tribute: One-third of all of their children would be brought to her, for her to kill and eat in a great banquet. The sisters had no choice. If one refused, she would be banished or killed, and all her children would be destroyed. Winter was coming, and it was too late to find a new castle or new suitors. The sisters watched as the new queen devoured every third child, her own nieces and nephews. The queen grew larger, and meaner, and never left the castle, not even for a minute.

One day the queen grew even more greedy. Not content with one child out of every three, she demanded an even greater tribute. Now she wanted one out of every *two* children brought to her for her gruesome repast. The queen's sister, the one closest to her in beauty and power, was furious. She chased the queen into her chambers, pinned her down, and bit her all over her face and body.

The queen, now duly chastened, stopped demanding such an onerous tribute. She would again be content with eating only one child out of three. The land was once again at peace.

I lied to you. This is not a fable. It's a true story, except that paper wasps, *Polistes fuscatus* ("smoky-winged city founders"), do not give birth to live children. After mating, they lay eggs that hatch into larvae within a few days. They do indeed live in social groups ruled by an alpha queen and multiple subordinate beta queens. The alpha dominates reproduction by eating about one-third of the eggs of her betas, who are her sisters. But if the alpha queen eats more than her "fair" share, the beta queens, particularly if they are close to her in body size, will punish her by biting her. That bitch can't get away with it.

Do paper wasps have a moral life? Morality is commonly held to be a uniquely human domain, but paper wasps, and indeed all sorts of nonhuman animals, react to social situations in eerily familiar ways. The paper wasp bites back when her sister gets too greedy. Worker bees attack, and sometimes kill, other bees who fail to do their fair share. The philosopher Peter Strawson described how the "reactive attitudes" of resentment, indignation, condemnation, and gratitude arise when people are "directly involved in transactions with each other." But human beings are not the only organisms who are in transaction with each other, who have social roles and expectations. Wasps in colonies, wolves in packs, crows in murders, humans in societies—members of all social species depend on each other to survive.

Cooperation emerged in social species, like wasps, long before there were humans, and we can follow the evolutionary trail of cooperation even further back, to the emergence of the first multicellular organisms, about a billion years ago. The multicellular body requires cooperation among the different cells, who sacrifice their own individual reproduction and mobility, and commit to specialized tasks and preprogrammed death. And just as there are greedy wasps, there are also cancer cells that steal more than their fair share of resources, abandon their work, trash their environment, and recklessly damage the health of the body collective. Their failure to cooperate, as I discussed earlier, can spur new innovations, but it also invites the body's discipline: The immune system works to contain and kill the seditious element.

Let's follow the problem of biological cooperation even further back: The individual cell contains genes that can replicate at the expense of the genome. Transposable elements, or "jumping genes," are DNA sequences that can behave selfishly, replicating themselves even as they impair the functioning and replication of the genome as a whole. The genome, in turn, has mechanisms to enforce cooperation by silencing selfish transposable elements.

Let's go even further back, to around four billion years ago, to the origin of life on earth, when chemistry somehow gave rise to biology. In the beginning was the word, and the word was an RNA molecule that could replicate itself. Even these primitive building blocks of life existed together in "communities of replicators," and molecular cooperation was essential for their survival, replication, and increasing complexity. Some molecules, for instance, only worked when they were assembled in a group and every member of the group was present. Others would replicate other strands but could only be replicated by another molecule. All of life, then, "is organized in a hierarchy of cooperation: genes work together in genomes, genomes in cells, cells in multicellular organisms, and multicellular organisms in eusocial groups."

Evolutionary theories propose that our moral lives—our resentment and gratitude, our blaming, shaming, and punishing, and even our praising and forgiving—emerged to help solve the problem of cooperation. Darwin first ventured this proposition in 1871 in *The Descent of Man, and Selection in Relation to Sex*, writing, "Any animal whatever, endowed with well-marked social instincts . . . would inevitably acquire a moral sense or conscience, as soon as its intellectual powers had become as well, or nearly as well developed, as in man." He hypothesized that the instincts he saw as evolutionary precursors to morality in nonhuman animals "were first developed, in order that those animals who would profit by living in society, should be induced to live together." Consistent with Darwin's original insight, primatologist Frans de Waal has documented how nonhuman primates show "building blocks of moral systems": They console each other, mediate social conflict, and retaliate against stingy animals who don't share food.

In humans, the dynamics of cooperation are sometimes studied in the laboratory, where researchers have the power to set up rules governing how people can interact with each other. In one such study, participants were asked to choose which of two groups they wanted to join. In both groups, participants interacted with each other in a "public goods game," in which each person was given an initial endowment and was asked to choose how much money they would contribute to a public good—the profits of which are returned to everyone equally—versus keep for their private account. In only one group, however, were people allowed to engage in costly punishment: They could pay to deduct money from other people whom they thought had made insufficient contributions to the public good. Initially, most participants chose to join the group that did not allow punishment. Over time, however, cooperation in the punishment-free group collapsed: Contributions to the public good were minimal, and profits to everyone plummeted. By the end, nearly everyone switched over to the group where they could punish and be punished. The authors concluded, based on their and others' results, that "the possibility of sanctioning norm violators stabilizes human cooperation at a high level, whereas cooperation typically collapses in the absence of sanctioning possibilities."

Is it any surprise, then, that our moral emotions are exquisitely responsive to failures of cooperation: defection, deceit, cheating, theft, lying, freeloading, betrayal? Human societies vary richly in their social and moral norms, but prohibitions against lying to or stealing from another member of your group are nearly universal. And the most serious violation of cooperation is deliberate harm to another member of your group. In Dante's *Inferno,* the deepest circle of hell is reserved for men who betrayed those nearest to them: Brutus, Cassius, Judas.

In turn, the most serious violations of cooperation are often met by yet more deliberate harm. The sister-wasp bites the greedy queen. The three-headed Satan gnaws on Brutus, Cassius, and Judas for eternity. The archway to Dante's hell proclaims the justice of retributive punishment: "Divine power made me, / wisdom supreme and primal love." Throughout the poem, the poet's sympathy for the damned is implied to be a sign of his inferior virtue. Visitors from heaven show no such compunction at the sinners' torment. Retributive aggression—an evolutionarily ancient impulse that we humans share with insects—is imagined not as bestial but as divine, not as primitive but as wise.

The 2020 movie *Promising Young Woman* stars Carey Mulligan as Cassie, a thirty-year-old woman who has dropped out of medical school and is living with her parents. She is obsessed with what happened to her best friend, Nina, who was raped by another medical student, Al, and who then committed suicide. Nina's allegations were never believed or investigated, either by the police or the school, and Al is now a doctor, engaged to be married. By day, Cassie works in a coffee shop. At night, she goes out, pretending to be drunk to the point of incapacitation. Men take her home and try to have sex with her, even though she is visibly incapable of consent. The climax is never the one the men are hoping for. Instead, the viewer's desire is gratified. Cassie, abruptly revealing her sobriety, confronts them: "What are you doing? What. Are. You. Doing."

As more evidence about Nina's rape is revealed, Cassie's plots thicken. The people who minimized Nina's anguish are her targets. Despite her girlish braid and macaron-colored clothes, she is reminiscent of the Furies, the three ancient Greek goddesses of vengeance, the daughters of hell and night.

The premise of the movie—a woman is raped and her rapist faces no consequences—is realistic. An estimated one out of six women in the United States and the United Kingdom have been the victim of rape or attempted rape in their lifetime. Most of these assaults are never reported to the police. It is estimated that less than 10 percent of rapists are ever incarcerated. When thinking about the harms suffered by victims of sexual assault, the desire for vengeance seems all too human.

The trailer advertises the movie as "a delicious new take on revenge." A spooky, spiky string arrangement of Britney Spears's "Toxic" plays in the background, an apt soundtrack for a film that looks like cotton candy and goes down like Drano. *I took a sip from my devil's cup. . . . You're dangerous, I'm loving it.* And we *are* loving it, at least at first, the fantasy of making wrongdoers suffer. I paid money to watch *Promising Young Woman.* I paid for pleasure, the dopamine kick of seeing ordinary white men, in their khakis and button-ups, look panicked, scared, helpless. I paid for the pleasure of seeing them squirm.

Retribution is indeed delicious. We can see that in the brain. Ordinarily, when people see someone hurt—receiving an electrical shock, for instance—they show an empathic neural response. Their anterior insula, an area of the brain that responds to pain, responds as if they are also being shocked. This reverberation of another's pain, as I've described, is crucial for child socialization. If, however, the person being shocked is first shown as betraying another person's trust, then the empathic response is muted or even totally absent. Instead of showing the neural signature of empathic pain, when people see the double-crossing traitor being shocked, their brains show activity characteristic of pleasure, with increased blood flow to the dopamine-rich nucleus accumbens area. The brain on retribution, in other words, looks like the brain on sex, food, or drugs. And as with sex, food, or drugs, people will pay for the dopaminergic reward of being able to administer punishment to a perceived wrongdoer. *I need a hit, baby, give me it.*

This appetite for retribution emerges early in life. Developmental psychologists Paul Bloom and Karen Wynn devised an ingenious set of studies showing that babies as young as eight months old can be punitive. First, babies were shown a puppet struggling to open a box. Next, another puppet appeared. The new puppet either helped open the box (the Helper) or sat on it, making it even harder to open (the Hinderer). Then the Helper or Hinderer puppets were shown playing with a ball that rolled away to a third puppet, who either gave the ball back (the Giver) or took it for themselves (the Taker). At this point infants were given the chance to interact with either the Giver or the Taker—which one did they reach for?

Nearly all the very young infants, who were just five months old, chose the Giver. They wanted the puppet who would give the ball back. But eight-month-old infants were more discerning: They preferred the Giver if it had returned the ball to the "good" Helper puppet, but they preferred the Taker if it had refused to return the ball to the "bad" Hinderer puppet. Slightly older children (nineteen-month-old toddlers) went even further. They didn't just prefer the Taker; they wanted to *be* the Taker. They showed a willingness to take a treat away from the Hinderer puppet. These studies show how early the desire for retribution emerges. Two wrongs do, in the eyes of a toddler, make a right.

Another study demonstrates how the pleasure of retribution emerges in childhood. In this study, children were first given an endowment of tokens, which could be redeemed for stickers. (Stickers, in case you haven't spent time with the kindergarten set recently, are a premium prize for this age group.) Next, they were shown a puppet, who either gave the children some of their favorite toys (which had been sneakily imported into the lab by mom or dad before the experiment) or taunted the children with the toys before snatching them away. Then an "assailant" puppet materialized, surprising the first puppet by hitting it with a stick as it cried out in pain. Mid-assault, the curtain to the puppet window closed. The kids then had the option: Would they like to pay a token to raise the curtain again and continue to see the first puppet be hit with a stick?

Four-year-olds were willing to spend their tokens freely, regardless of the puppet's earlier behavior. They wanted to watch whatever was available. Six-year-olds, in contrast, were more discerning: They would pay to watch, but only if they were watching the wrongdoing puppet being assailed. They also smiled as they watched. And in a twist that Darwin would have appreciated, the researchers also did a version of the study using chimpanzees instead of children. Just like the six-year-old kids, chimpanzees preferred watching the wrongdoer be punished.

We get dopamine hits from retribution; we will pay to watch retribution. These results are not surprising from an evolutionary perspective. Retribution is an ancient, evolved mechanism for enforcing cooperation. Wasps will bite each other for being greedy; fig trees, which depend on wasps for pollination, will abort figs containing wasp eggs if the wasps have been lazy in their pollination. But administering punishment can be costly; it takes effort and can invite retaliation. If there are costs to a behavior, pleasure can be its recompense. The ultimate evolutionary explanation for why some behaviors are rewarding is that organisms who felt reward when they did those behaviors did them more often, and those who did those behaviors more often lived longer, mated more, kept their offspring alive longer.

Our desire to punish, then, is no less tethered to our animal past than the desires that are punished. But retributive desires—all evolved desires—are "blind to the forces that explain their existence," bubbling up in situations divorced from their ultimate adaptive end. Just as people experience sexual pleasure in all sorts of situations in which no babies will be made, people experience punitive pleasure even when no cooperative norms will be maintained by the retributive act. Punishment might not deter future defections or prevent future harms; it might even make them more likely. And we want it anyway.

Nietzsche, an astute observer of human nature, anticipated all this. Long before social neuroscience, he knew that punishment was a lust. One clue: our curious habit of talking about punishment in terms of debt and repayment. How, by suffering, does one repay one's "debt" to society? Why, in German, does the word *Schuld* (related to the English "should") mean both "debt" and "guilt"? Why do we have such a similar lexicon for our financial lives and our moral lives? *Accountable. Duty. Fault. Forgiveness. Reckoning. Reconcile. Responsible. Scot-free.*

The answer, according to Nietzsche, is that cruelty is a currency. The wrongdoer injured you in some way; he detracted something from your life. He is now in your debt and must give you something valuable in return. What he is repaying you with is the pleasure you derive from watching him hurt. When an offender "pays his debt to society," society is being repaid with the indulgence of an ancient, evolved desire to hurt.

The strength of the punitive desire differs between people. Punitiveness is part of a constellation of personality traits that psychologists label "authoritarianism," defined as a "desire for group conformity at the expense of personal autonomy, accompanied by a deference to in-group authority figures and a desire to punish those who violate in-group norms." (Possibly because psychologists tend to lean left politically, most research has focused specifically on right-wing authoritarianism, but left-wing authoritarianism is now an emerging subject of study.) On average, authoritarian individuals tend to be high in self-control. They are less interested in art, imagination, variety, and unusual experiences. They like closure, dislike ambiguity. And they want to punish more severely. Authoritarian individuals are more likely to say that torture is justified. They are more likely to support capital punishment. They are more likely to spank their children.

Authoritarian punitiveness is as stable as other aspects of personality. One study followed people for fifteen years and found that levels of authoritarianism were highly stable from the beginning to the end of the study period. Just as adults rarely become radically more introverted or conscientious in adulthood, neither do they become radically more or less authoritarian. And authoritarianism is as heritable as other personality traits, with about half of the variation between people in their authoritarianism due to genetic differences.

The personality trait of authoritarianism is vividly illustrated in the comic book character called the Punisher. Marvel introduced the Punisher as a character in *The Amazing Spider-Man* in 1974. The character's backstory was this: Frank Castle, a native of Queens, New York, was planning to become a priest until he realized that "there could be no forgiveness without punishment." Instead, he joined the Marines. After serving in the Vietnam War, Castle returned to the United States, where his wife and children were murdered by mobsters. Unable to get justice for his family due to police corruption, Castle turned to vigilantism, waging a war on vice.

The Punisher has become a ubiquitous emblem among military, police, and far-right paramilitary organizations and their supporters. In his memoir *American Sniper*, Chris Kyle, a Navy SEAL who served in the Iraq War, wrote of his unit's fondness for the Punisher logo: "We spray-painted it on our Hummers and body armor, and our helmets and all our guns. And we spray-painted it on every building or wall we could." (Back in the United States, Kyle was murdered with his own weapon by another veteran of the Iraq War who had served as a military prison guard.) During the January 6 insurrection, when more than two thousand supporters of President Donald Trump invaded the U.S. Capitol, the Punisher skull could be seen everywhere: stickers, T-shirts, flags, tactical vests. In 2023, the New York City Police Department was obliged to remove a Punisher poster from its Sex Offender Monitoring unit. The poster proclaimed: "I hunt the evil you pretend doesn't exist."

The Punisher character is represented by an elongated white skull, strikingly reminiscent of the white skull patches that festooned the uniforms of the SS units—the "Death's Head" units—that administered the Nazi death camps. The *Totenkopf*, for the Nazis, was the perfect icon for authoritarian values, which prize cooperation and conformity for the sake of the in-group above the potentially rebellious dictates of the individual conscience. As Heinrich Himmler, one of the key architects of the Holocaust, put it: "The skull is a reminder that you shall always be willing to put yourself at the stake for the life of the whole community."

The Punisher embodies a moral ambiguity. In the words of one commentator, "Is he a troubled good guy or a fascist murderer?" The same ambiguity hangs over all retributive violence. Is retribution a form of righteous justice, or merely an excuse to indulge in cruelty? Anthropologists Alan Fiske and Tage Rai argued, in their book *Virtuous Violence*, that much human violence is motivated, as the Punisher is motivated, not by amorality or psychopathy or impulsivity but by a sense of moral outrage: *They can't just get away with it.*

Where do you feel outrage in your body?

I was raised a good Christian girl; I am a millennial eldest daughter. I was a Pleasure to Have in Class. Which is to say, it's difficult for me to identify outrage in my body. A classic Freudian neurotic, who represses anger, intellectualizes it, turns it against the self. My therapist regularly, gently prods: Are you sure you're sad? Could you possibly be . . . angry?

When I let myself feel angry, I feel like I'm choking. My ribs flare; my solar plexus hums. My cheeks are hot, and the soles of my feet tingle. Words tumble out, so many words, imaginary conversations that loop in giant circles, returning, again and again, to the same spot. I cry when I'm mad, fat toddler tears of frustration that the world is different than I would have it. And, then, back to numbness. A callous, contemptuous remove that I feel behind my eyes and at the base of my neck.

As I work on this book, I text with a friend whose husband first started choking her when she was three months pregnant with their first child. That was more than a decade ago. He is now on a work trip, and she fantasizes that his plane crashes on the way home. She doesn't want the plane crash to be quick, or painless. She wants flames. She wants anguish.

I can hypothesize about why her husband behaves as he does. Might he have inherited a constitutional predisposition toward disagreeableness? Maybe. Was he taught from a young age by the adults in his life that explosive male anger was acceptable? Probably. But does knowledge of the why temper my, or her, rage at the what? Not really. Nor do I find much consolation in philosophy. My reactive attitudes are, for a time, beyond the reach of arguments about determinism. What do these academic word games have to do with choking? The question of why he behaves as he does, in general, is less salient to me than my outrage at how he behaves as he does *to her*. Is he just going to get away with treating her like a beta bitch who can be pushed around?

My own family resentments are more ice than fire, more numbness than anger. Within a few months of discovering the refrigerator postcard calling me "unwanted undesirable," I stopped communicating with my mother entirely, joining the nearly 30 percent of Americans, and 20 percent of British adults, who are estranged from a family member. Many adults who have gone "no contact" with their parents also grew up in highly religious and authoritarian families. Unlike my friend in an abusive marriage, I do not fantasize about flames of retributive punishment. I do not want my mother to suffer. I simply cannot bear to be "directly involved in transactions" with her anymore. I have responded to the queen's aggression by abandoning the nest.

In the *Oresteia*, a trilogy of plays written by Aeschylus, a wife's anger at her husband, and a child's at his mother, eventually lead to the first trial by a jury of twelve citizens. Like evolutionary theories, the plays describe the family history of justice, rooting the abstractions of law and philosophy in our intimate relationships.

The drama begins when Agamemnon's army is stranded in the harbor, unable to sail to Troy to recapture the fair Helen. Agamemnon learns that the only way to appease the gods is to sacrifice his daughter, Iphigenia—the violence he gives Necessity's name. Many years later, when Agamemnon returns victorious from Troy, his wife, Clytemnestra, avenges her daughter by stabbing her husband in the bathtub. Then their son, Orestes, avenges his father by killing Clytemnestra. Then the Furies—with bat wings, and dog heads, and mouths full of undigested blood—hunt Orestes. *Lex talionis,* the law of retaliation, perpetuates a vicious cycle of bloodshed, which ends only with the goddess Athena's intervention. She puts Orestes on trial, choosing twelve Athenians to serve as the jury, and convinces the Furies to live, transformed and renamed, in a cave under the city of Athens. Private vengeance is sublimated to public justice.

I find it notable that the Furies are persuaded, constrained, repressed—but not eliminated. The three sisters live. Perhaps they could not be eliminated. They are, after all, older than the Olympian gods. Athena respects them as a primordial force. She makes a place for them. Aeschylus suggests a continuity between the primitiveness of retributive energy and innovations in justice. Retributive instincts cannot be expunged but can be transformed.

In contrast, some scientists and philosophers have argued that the Furies should be destroyed, not just contained. The biologist Robert Sapolsky, for instance, argued that we should view other people as if they are inanimate forces of nature—as *dangerous,* perhaps, but not *bad.* To want to punish wrongdoers is, in his view, to be as ignorant as "goitered peasants who thought Satan caused seizures." Similarly, the philosopher Derk Pereboom has argued that resentment, indignation, and outrage toward people who have harmed us or others should be resisted and limited as much as possible. The recommended course of action, when someone is experiencing or expressing "inappropriate" moral emotions, is "ignoring them [the emotions], prodding the agent to reflect further on them, giving him or her space to get over them, encouraging him to talk about them in suitable ways." In other words, someone who insists on morally condemning another should be treated like a toddler having a tantrum.

Even as I recognize something disturbingly cruel in retributive urges, there also seems to be something cold-blooded, not to mention condescending, in the suggestion that outrage at a harmful wrongdoer is simply a matter of being insufficiently rational or scientifically ignorant. Yes, our retributive sentiments are deeply rooted and deeply dangerous. Yes, we can delight in harming wrongdoers and are tempted to invent classes of wrongdoers so that we can revel in harming them. And, yes, our behavior is caused, just as the storming sky is caused, just as the viral infection is caused. But does that mean that hating anyone for anything they've done is absurd? Has science rendered moral outrage obsolete?

In his essay "Freedom and Resentment," the philosopher Peter Strawson answers those who think that knowledge about the determinants of our behavior renders our moral practices moot. Crucially, Strawson does *not* argue that human behavior isn't determined; he is not looking for a metaphysical loophole where freedom might be hiding. He doesn't talk about the differences between identical twins, or the limits of statistical prediction, or the indeterminism of subatomic particles. Instead, he argues that determinism is, in the end, simply irrelevant to the question of whether we are morally responsible, because responsibility does not depend on some exemption from nature. In Strawson's view, to be morally responsible is, fundamentally, to be a person in human relationships—to be subject to their demands and expectations. Moral responsibility is a social condition, not a supernatural one.

And because human beings necessarily exist in societies, the practice of holding each other morally responsible is an inescapable feature of being human. The philosopher Pamela Hieronymi, in her book on Strawson's "social naturalism," summarizes it this way: "The existence of some such framework of interpersonal demands and reactions is a natural part of human life—a part that can be improved and refined, to be sure, and, can even be improved and refined for moral reasons, in light of our increased understanding of our place in nature—but not part of life that could or even should be abandoned."

One way to understand this argument is to think of how blaming and punishing are like another natural part of human life—sex. Sex, like punishing, feels good; it activates our reward systems. Sex, like punishing, is part of humanity's deep evolutionary history. And sometimes sex, like punishing, is something that philosophers say we should abstain from entirely. Augustine thought that sexual desire was proof of God's punishment; he chose celibacy for himself and urged it on his followers. Kant worried that sex necessarily involved treating another person as a means, rather than as an end in themselves, and could only be acceptable if done for the sake of conceiving a child in marriage.

The problem, of course, is that sexual reproduction is an inescapable feature of the human species. We are mammals, after all. Some people are celibate, of course, but humanity without sex altogether would no longer exist. Sexual reproduction emerged two billion years ago with the exchange of genetic material between single-celled protists, and over 99 percent of eukaryotes—ranging from bacteria to hamsters to people—reproduce sexually. There is no rewinding the evolutionary clock to achieve the Augustinian fantasy of sexless humanity; the concept is incoherent.

Strawson applied this same logic to moral responsibility. Like sex, blaming and punishing and holding responsible are deeply baked into human nature. And just as humanity cannot exist entirely without sexual reproduction, humanity, he argued, cannot exist without a system of interpersonal demands and expectations and corresponding moral attitudes, including condemnation and blame. What if our behavior is truly determined by factors, including our genes, beyond our control? Strawson: "Such a question could seem real only to one who had utterly failed to grasp . . . the fact of our natural human commitment to ordinary interpersonal attitudes. This commitment is part of the general framework of human life."

In other words, even if determinism is true, humans cannot just entirely give up on holding each other morally responsible, of *mattering* to each other in a particular way, any more than we can entirely give up being mammals who sexually reproduce. If "humanity without reactive attitudes" is an incoherent concept, then it is incoherent to argue that something universally true of all humans—like our behavior being embedded in the pure cause and effect of the animal world—demands that we entirely abandon our all-too-human practice of praising each other, blaming each other, holding each other responsible.

At first, this line of reasoning seems to lead back to the same horrific thought that tormented Travis in the desert: I will hold people responsible for what they do, even though I know their actions are shaped—perhaps even determined—by forces beyond their control, and I, in turn, will be held responsible. This conclusion does, indeed, echo the doctrine of original sin, with its grim insistence that humans are on the hook for their behavior despite its ineluctability. Except—the hook is not divine, piercing us in a lake of fire. The only hooks we have are the ones we fashion for each other. Blame and punishment are, like sex, something that happens in this life, something we do *with* our fellow creatures. We blame and punish, not in spite of being part of the animal world, but because we are.

Hooks: In Dante's *Inferno,* the demons in hell punish wrongdoers with hooks. "Each one of these sinners must be under the pitch; / If not, the hook must catch him and keep him down." In everyday English, "on the hook" can mean being physically trapped or financially obligated; a fish, literal or metaphorical, is impaled by a consequence of its appetite. But there are other types of hooks. The vaudeville hook pulls someone back, physically enforcing a social contract. And the crochet hook grabs a strand of yarn, weaving it together with others, creating a whole that is stronger and more flexible than any individual thread.

We like to believe that we will be blamed for our choices, that we won't be found guilty unless we had the opportunity to make different choices, that we can avoid damnation by making better choices. That we won't wriggle on the hook unless we had a chance not to take the bait. I have come to believe, though, that all of our choices are structured by much we didn't choose—and that we can be guilty anyway. But, crucially, guilt does not mean damnation, and the hooks of moral responsibility can be a tool for weaving, and reweaving, us together.

Hooks: In 2012, Adam Lanza, a gaunt and reclusive young man, murdered his mother, twenty children and six schoolteachers at Sandy Hook Elementary School, and then himself. The inevitable questions of nature and nurture were asked: Was his violence caused by an autism spectrum disorder, or undiagnosed schizophrenia, or some other genetically caused mental disorder? Or was he perhaps a victim of undiscovered child abuse? "Chilling Look at Newtown Killer, but No 'Why,'" read one *New York Times* headline. The world asked of Lanza the same questions my prison correspondent was asking me: Why? What would drive someone to do such a thing? What makes a child go bad?

I would never presume to tell Sandy Hook parents that they were acting like "goitered peasants" if they hated the man who slaughtered their babies. But they, acting with a fortitude I can describe only as holy, have not fixated their condemnation solely on Lanza, nor on his parents. Their outrage and blame radiate to the gun manufacturers who profit from carnage, the lawmakers who allow young men ready access to instruments of mass death, and an American culture that is willing to sacrifice its children to the demonic idol known as gun rights. When a textile is rent, repairing it requires attending to more than a single strand of yarn. Hooks can be used to weave a new cooperative peace.

In *The Reckonings*, the writer Lacy Johnson considers what she wants from the man who kidnapped her, raped her, tried to kill her. He is not in prison. He is living free, married, raising two children. She shocks her audience, who express bloodlust on her behalf, by separating reckoning from revenge. "I don't want him dead. I don't even want him to suffer." Her response emulates the biblical God, who did not kill Cain for murdering Abel, or even wound him, who permitted him to live and father children and found a city. She writes: "More pain creates more sorrow, sometimes generations of sorrow, and it amplifies injustice rather than cancels it out." Her vision for reckoning involves his admission of guilt and his submission to community: "I want him to admit all the things he did, to my face, in public, and then to spend the rest of his life in service to other people's joy." She will not get this. But the futility of her hope does not destroy the wisdom to be gleaned from listening to her conception of justice.

The philosopher Jean Hampton, in her book *Forgiveness and Mercy*, gives a compelling account of what type of retributive urge is being expressed here. Retribution, in Hampton's view, properly aims not at violating the wrongdoer, not at making them suffer, but at defeating them to communicate a truth about equality of human value. Hampton writes:

> Those who wrong others . . . objectively demean them. They incorrectly believe or else fail to realize that others' value rules out the treatment their actions have accorded the others, and they incorrectly believe or implicitly assume that their own value is high enough to make this treatment permissible. . . . By victimizing me, the wrongdoer has declared himself elevated with respect to me, acting as a superior who is permitted to use me for his purposes. A false moral claim has been made. Moral reality has been denied. The retributivist demands that the false claim be corrected. The lord must be humbled to show that he is not the lord of the victim.

We can hear this idea, that retribution aims to humble, in the English phrase "brought down." An article about Harvey Weinstein's conviction for rape: "Allegations of sexual assault dating back decades brought down one of the most powerful producers in Hollywood." We can also hear the idea that such humbling affirms the worth of the victim. After the Minnesota Court of Appeals affirmed the conviction of Derek Chauvin for the second-degree murder of George Floyd, the prosecutor on the case, Attorney General Keith Ellison, commented: "The Court's decision today shows once again no one is above the law—*and no one is beneath it.*" Not to punish Chauvin, he is saying, would be to communicate something morally wrong about the relative value of George Floyd, an unarmed Black man, and Chauvin, a white police officer. Chauvin's conviction was widely celebrated as an affirmation of Floyd's value—an affirmation that Black lives matter. Such an affirmation was not a foregone conclusion in a racist society that once permitted Black people to be owned and harmed with impunity. (*Impunity:* the state of not being punished.) If punishment communicates a society's understanding of the value of human lives, of who is and is not allowed to use whom for what purposes, then people who have been devalued and demeaned will, rightly, demand punishment for those who wrong members of their community as a way of asserting their equality of worth.

Notably, defeat does not necessarily involve causing someone to suffer, although, obviously, inflicting suffering is a common instrument of defeat. Consider, for instance, when offenders are sentenced to community service. Delivering meals to housebound older adults, picking up trash on the side of the road, walking dogs for the Humane Society—these experiences need not be painful, and one hopes that the offender grows to find meaning and value in contributing to the larger good. But painful or no, these acts are *service*—that is, labor that is traditionally performed by people who occupy the lower ranks of the social hierarchy, labor that benefits the group more than the individual performing it. Serving others, therefore, rectifies the faux superiority communicated by the original wrongdoing.

Hampton's analysis is consistent with how other cooperative social species commonly use punishment to establish and maintain dominance relationships. Generally, dominant animals are more likely to punish subordinate ones for acts that threaten cooperation or the fitness of the dominant individual, than vice versa, but subordinates will retaliate against dominant individuals when the opportunities are ripe, and particularly when they have overstepped their roles. To punish, then, is to communicate something about their status. The paper wasp who attacks her excessively egg-eating sister, for instance, is using the language by which social orders were originally formed to communicate something about the violation of that order: The queen made herself too high, took too much, so now she must be brought low. The form and function of retribution in social species can never be separated from questions of power.

Hampton's analysis is also consistent with cross-cultural and evolutionary work on the moral emotions, including guilt and shame. In some cultures, the same word is used both for shameful, transgressive experiences and for being in the presence of a person who ranks higher in the social hierarchy. For instance, anthropologist Daniel Fessler wrote that inhabitants of a semi-traditional fishing village in Indonesia used the word *malu* for the emotion they would feel if they ate during a fasting period, indulged in alcohol or sexual pleasure, failed to visit their sick neighbors—and if they were a young person asking something of the village head. This linguistic overlap suggests that "the key to understanding the moral nature of shame" might be "to recognize that it has a phylogenetically older and simpler version," which is the emotion that accompanies being downranked in a dominance hierarchy. Looking beyond humans, nonhuman mammals make facial expressions and bodily postures when they are displaying submission to a higher-ranked animal that are very like what humans do when they are expressing guilt, shame, and embarrassment: downcast eyes, shrinking body size, an incongruous smile.

The relationship between human expressions of guilt and shame and nonhuman displays of submission helps illuminate why we want someone to express remorse after a transgression, and why it feels so galling when they don't: The wrongdoer is communicating that he has not been defeated. He still considers himself to be superior to the victim, whom he used as if she were inferior. Lack of remorse, then, is not just adding insult to injury;

rather, it compounds the original injury, which involves a status injury. "Who do you think you are?" we ask the offender. What we are really asking, though, is, "Who do you think I am, that you treat me like this?"

The desire to see a wrongdoer defeated has its limits. If we are asserting, with our retribution, the equal value of every person, then that includes the offender as well as the victim. The wrongdoer ought to be humbled, not humiliated. Defeat ought not entail degradation to a condition beneath basic human dignity. Christian depictions of hell, where people's entrails are chewed on for all eternity, where people walk around with their heads on backward—this is debasement, not defeat. So, too, are the conditions of many prisons, where people are deprived of clean water, nutritious food, basic physical and mental health care; are victimized; are sometimes held in sanity-shattering isolation. This is torture porn. Retribution, in this account, appropriately aims to affirm the equality of wrongdoer and victim, not to use the "deservingness" of the wrongdoer as an excuse to rob them of human dignity.

I have come to see punishment, then, not as an antidote to our creatureliness but as another manifestation of it. It remains a dangerous but indispensable part of human life—one that can be used to degrade or express the equality of human relations.

Not just punishment; forgiveness, too. The contemporary American poet Carl Phillips wrote, "It's as if forgiveness were, in fact, an animal—wild / like animals, the particular wild of animals that have / lived domesticated their entire lives." I interpret these lines to mean that as much as we would like to pretend that our forgiving and punishing can be fully tamed—made orderly, rational, sanitized; made to come or go on command—they remain, like our sexuality, bodily experiences that reveal their primal history.

And if the Furies were sometimes portrayed as feral dogs, I imagine forgiveness as feline: beautiful, fickle, unbiddable. Perhaps it will draw close to us, soothing and warm, but it will not come just because we command it. Especially not if we haven't cultivated a relationship with forgiveness, gained its trust. I used to interpret the Lord's Prayer—*Forgive us our trespasses, as we forgive those who trespass against us*—as a bargain I was making with God. I will force myself to forgive, and in return, God might forgive me. Now, though, I interpret both halves of this line as a plea to a force more ancient than the Christian conception of God. Please, forgiveness, come to me.

In our personal relationships, forgiveness can be invited but rarely commanded. In our institutions, especially in our criminal justice systems, we have more room for deliberate choice: punishment or forgiveness? Drawing on evolutionary theory, the philosophers Nicola Lacey and Hanna Pickard point out that *both* are adapted strategies for responding to harm. To err is animal, and to forgive is animal, too. The sister wasp will bite the excessively egg-eating sister but will also refrain from hurting her permanently, allowing the possibility of their social relationship to resume. And, of the two evolved strategies, forgiveness has a higher upside. "From an evolutionary perspective," they write, "reparation as opposed to retaliation is optimal when successful, for it reduces the risk for the exploiter perpetrating future harm, without incurring the costs of monitoring and maintaining the power to retaliate, and while bringing the benefit of preserving relationships so far as possible." In other words, we should strive for forgiveness for the sake of the wrongdoer and for our own sake.

Given the advantages of reparation over vengeance, Lacey and Picard go so far as to suggest that criminal punishment should be "reconceived as an institutionalized form of forgiving." Not the abolition of punishment, but punishment *with* forgiveness! In practical terms, this approach might involve not condemning an offender's character as inherently deviant or defective; clearly communicating to the offender the damage done to victims; providing offenders a voice in determining their own sentencing, including options for public service, therapy, and continuing education and workforce training; and facilitating offenders' close relationships with significant others with more regular family visits and part-time sentences. These suggestions would be a radical departure for the United States criminal justice system, which was founded on the Christian ideas that vengeance is divine and that forgiveness can only be claimed after a blood price has been paid. But if we reject those ideas, if we embrace our shared dilemma of being animals who—in Darwin's words—would profit by living in society, if we recognize our shared evolutionary heritage as creatures who both desire to punish and long to forgive, then new possibilities emerge.

As someone who was taught from a tender age that divine justice meant perpetual torture, it can be hard for me to imagine what alternate forms of accountability and forgiveness look like in practice. But as we will see, societies can forswear the most virulent forms of retribution but still insist on holding each other responsible. What happens when we don't try to abandon the practice of holding each other morally responsible but do temper the lust for inflicting hell?

Eclipse

In which I witness a solar eclipse and wonder at the dualities of being human: We are children and adults, victims and perpetrators, at the mercy of forces bigger than us and needful of each other's mercy

LET ME TELL YOU one last fable.

Once upon a time, when there were still giants on the earth, God saw that humans had become very wicked. Every imagination of their hearts was evil. All flesh was corrupted. He repented of having created humans and vowed to destroy them, along with the beasts of the field, and the fowls of the air, and all the creeping things.

Because he was convinced of their wickedness, God drowned everyone. Only one family, who built a large boat in anticipation of the end of the world, survived, along with their animals. It rained for forty days and forty nights. All in whose nostrils was the breath of life, all that was on the dry land, died.

Andrea Yates was convinced that children were evil from their youth, but she also believed that if they died young enough, "God would take them up" anyway. So, to save her five children from hell, she drowned them. She then called 911 and asked the operator to send the police. "I need 'em to come." The dispatcher asked whether she was ill. "Yes, I'm ill," Yates replied. When police arrived, they found four dead children—ages five, three, two, and the baby, just six months—under a blanket on the bed. The oldest child, age seven, was found floating face down in the tub, a chunk of his mother's hair in his fist. His wet footprints, evidence of his attempt to escape, could still be seen on the bathroom floor. Yates was sitting on the couch, minimally responsive, soaked.

No one contested that Yates was seriously mentally ill. Before she murdered her children in 2001, she had already been hospitalized four times for suicidality, delusional thinking, and intrusive thoughts and had been diagnosed with postpartum depression and postpartum psychosis. She was prescribed Haldol, an antipsychotic medication that blocks dopamine signals in the brain, after she threatened to cut her throat with a kitchen knife. She feared hurting her children. The psychiatrist who diagnosed Yates warned her husband that she was very sick and that another child would likely cause another severe psychotic reaction. They had a fifth, Mary, their only girl, the next year.

What was contested at Yates's trial was whether her mental illness made her not guilty, by reason of insanity, of murdering her children. The insanity defense is rarely attempted and is usually unsuccessful. The crux of the insanity defense is that the person, at the time they were committing the criminal act, didn't know what they were doing, or didn't know that it was wrong, or couldn't stop themselves from doing it.

Yates had rarely been left alone. She waited to draw the bathtub until her husband left for work, when she had one unsupervised hour before her mother-in-law was due to come and check on her. She immediately called 911 when she was done drowning her children. Prosecutors argued that her planning and her immediate summoning of the police were proof that she knew right from wrong and was in control of her actions. Yates's defense argued that, even if she knew that drowning her children was *legally* wrong, she had been convinced that drowning them was still morally right. The only way to save her babies from suffering in hell for their original sin was to make sure they died young enough for God to take mercy on them. At the time of the murders, her lawyers said, the fatty goop in her skull was malfunctioning from too much dopamine, too many hormonal changes, too much hell-mongering, all of which rendered her unable to understand the reality of right from wrong.

The jury at Yates's first trial disagreed with this assessment. They found her guilty of murder. But the first verdict was overturned because an expert witness for the prosecution gave false testimony, describing an episode of *Law and Order* that never existed. The jury at her second trial found Yates not guilty by reason of insanity.

The insanity defense presumes a clean divide among people. On the one hand, there are the insane, who are not responsible for what they do. On the other hand, there is everyone else, the sane, who are rational and reasonable, who know what they are doing, know that it is wrong, do it volitionally, and could have done otherwise. Cases like that of Andrea Yates, where the insanity defense is contested, where reason and planning are evident but so is delusion, complicate this neat divide. Two juries, faced with the same person and the same facts, reached different conclusions.

What we know about the hallucinating brain also complicates that neat divide. Psychotic-type experiences are much more common than people realize: Up to 8 percent of the population will, at some point, have one. Stress, sleep deprivation, and even some over-the-counter drugs like Benadryl can cause psychotic experiences, and cultures differ in whether they consider hearing voices abnormal. And psychosis itself—perceiving things that aren't really there, believing things that aren't really true—exists on a continuum with everyday perception. No one, ever, is perceiving the world as it really is. Our brains are stuck in the dark cave of our skulls, and they are inventing a phenomenology of experience. Ordinary perception, which is to say ordinary life, is a controlled hallucination, controlled in the sense that it is constrained by sensory input coming from the world outside our skulls. Hallucination is less-controlled perception. The boundary we draw between psychosis and normal perception is like the boundary we draw between green and blue, between child and adult, between grains of sand and a heap of sand. And the commonplace experience of hallucination reminds us that perceptual experiences can feel very, very real without being in any way veridical.

What if, some neuroscientists have suggested, the experience of a volitional self is best understood as a controlled hallucination? We live our lives perceiving ourselves to be a *self,* an "I," an ego, an executive, the wizard behind the curtain, a first-person player, a chooser. Big choices: Should I stay married? Small choices, like the one I made ten minutes ago: Should I clean up my toddler's high-chair tray or ignore the dirty dishes and sit down at my computer instead? (I chose the latter.) The small choices add up to the big ones, and the big ones shape the small ones. And throughout all of the choices is the perception of I-as-agent. But what if the perception of *being able to choose* is, at least sometimes, as manufactured by our brains as the perception I had on acid once, when I felt like I was able to flick the aspen trees off the mountain tree line with my fingertips? What if our belief in the volitional "I" is as manufactured by our brains as Yates's belief in hell?

That might sound insane, if your perception of a volitional self has never gone staticky or sideways. But mine has—on acid, yes, and in church, and in bed with a lover, and in the labor and delivery room. Many of my most profound moments, the moments that I identify in retrospect as pivot points, when I did or discovered something that altered my life, were moments when my perception of myself as a rational chooser dissipated. What remained in those moments was not pure sphexishness or base impulsivity. What was lost, when my ego was dissolved, was not my humanity. Complex human behavior and varieties of human experience worth wanting do not, in my personal experience, always and in all circumstances require the perceptual experience of a volitional self.

Back when I worked in an inpatient psychiatric hospital, fresh out of graduate school, many of my patients were suffering from active hallucinations. Antipsychotic medications, like the Haldol that Yates took after she gave birth for the fifth and final time, don't always get rid of symptoms entirely. I'm a psychologist, not a psychiatrist; my job wasn't to tinker with patients' dopamine levels by changing their drugs. My job, at least in part, was to help change their relationship to their perceptions of the world. One strategy in this type of therapy is to track changes in symptoms over time. By noticing how their hallucinations ebbed and flowed with sleep, medication, drugs, and stress, patients were encouraged to consider their hallucinations as something to be looked at rather than just lived through. "*I perceive* bugs crawling all over my skin" instead of "*There are* bugs crawling all over my skin." Patients might not be fully convinced that their brains are creating experiences that aren't true, but they are invited to consider the idea. They are invited to entertain doubt.

This is where I am now: I entertain doubt. I entertain doubt that my perceptual experience of rationally choosing what to do is always telling me something true about the wellsprings of my behavior.

Many people imagine this to be a dangerously subversive idea, because they consider the reality of the volitional self to be necessary for us to hold each other accountable for anything. I disagree. I think "holding accountable" is as basic to how humans relate to each other as sexuality. The rules for both are endlessly, culturally plastic, but no human society exists without both; indeed, the very notion of human society without humans mattering to each other in a way that we call "moral" is incoherent.

To be clear: I am not saying there are no morally relevant differences between the actively psychotic and the nonpsychotic. But in our current practices, we *already* hold people who have been deemed "not a rational chooser" accountable and set boundaries to prevent them from doing harm again.

The criminal justice system did not respond to Yates's murder of her five children with a collective shrug. She was arrested for murder; her act was publicly condemned as wrong. She will likely spend the rest of her life in Kerrville State Hospital, a low-security mental hospital just up the road from where I sit. Her liberty, already compromised internally, has been severely limited externally as well. What she was spared was the additional suffering and indignity inherent in a U.S. prison sentence. What she was spared was an even worse cage. What she did was horrifically wrong, and perhaps, given her circumstances and her brain, she could not have done otherwise. She is "not guilty," and she has been held responsible. She harmed others, and she needs care. And, and, and.

Her ex-husband wept when the not-guilty verdict came down. He appeared to be weeping with relief at Andrea being spared harsher punishment, but I wonder if he was also weeping with relief, or regret, for himself. He had, after all, impregnated a woman who was obviously, desperately sick. He had left a psychotic, grieving, self-mutilating woman alone with five young children in her care. I try to imagine what that moment was like for Andrea Yates. To know that you have murdered your children and then to see the father of your dead children cry with relief when you are spared. The agony, and the small mercy.

Legal mercy, even for people with psychotic disorders, is rare. Yates, despite her obvious psychiatric morbidity, was found "not guilty by reason of insanity" only after a second trial. Many people who are as ill as Yates are arrested, jailed, and imprisoned, a testament to the savagery of the modern punitive drive. One study of low-income men in the United States who had been hospitalized for a psychotic episode for the first time found that over half of them had been previously incarcerated. Another study found that 15 percent of state prisoners and 24 percent of jail inmates in the United States had psychotic disorder symptoms. Marcia Powell suffered from schizophrenia and was taking Haldol, which increases sensitivity to heat and sun, when she died in an outdoor prison in Arizona.

No one can seriously consider psychotic disorders to be a matter of "free will," however narrowly or expansively one interprets the term. Yet many people who experience psychosis are punished as if all their actions are as free as my choice to ignore the toddler's dinner tray. We assume our subjective perceptions of freedom are definitive about others' reality, even as those others show us how tenuous the connection between perception and reality can be. We justify punishment with appeal to choice: *The miscreant deserves punishment because he might have acted otherwise.* But the harsh treatment of people with psychotic experiences in the criminal justice system suggests a darker dynamic: The belief in the volitional self is sometimes a consequence of the desire to punish, not its antecedent. "The miscreant must have been able to act otherwise," we insist, "because she will be punished."

In the Genesis story, water covered the earth for 150 days and then abated. And God vowed never again to drown his creatures. He promised himself: "I will never again curse the ground because of humankind, for the inclination of the human heart is evil from youth."

That's the translation of Genesis 8:21 in the New Oxford Annotated Bible. The word after the comma is "for": Punishment will be restrained *because* the human inclination to evil can be traced back to our earliest days. The developmental roots of character mitigate blameworthiness. But some versions of the Bible, Jewish and Christian, translate this verse using "even though" instead of "for." The New International Version translation of this verse, for instance, is: "Never again will I curse the ground because of humans, *even though* every inclination of the human heart is evil from childhood."

Which is it? Are we to be given mercy *because* we can't help ourselves? Or are we to be given mercy *even though* we can't help ourselves?

Biblical scholars who argue for "even though" point to an earlier passage in the flood story, when God first decides to annihilate everyone: "And God saw that the earth was corrupt; for all flesh had corrupted its ways upon the earth. And God said to

Noah, 'I have determined to make an end of all flesh, for the earth is filled with violence because of them.'" It doesn't make sense, these commentators argue, for our corrupted flesh to be the cause of God's wrath before the flood covers the earth but mitigating afterward. God seems to change his mind.

If you read the Bible as literature, though, then the character of God obviously changes his mind. He spares Cain, even though Abel's blood cries up from the ground. Then he drowns nearly everyone in disgust. And then he repents of his vengeance. I think God's reversals and contradictions express *our* dilemma. This is the reading that forensic psychiatrists Paul Appelbaum and Azgad Gold offer in their article "The Inclination to Evil and the Punishment of Crime—from the Bible to Behavioral Genetics." They write, "If as the Bible suggests, God Himself appears to vacillate between the two alternatives—it is not surprising that human beings have such a difficult time with these judgments."

Vacillation can express a dialectic. Unable to hold two apparently contradictory truths at the same time, unable to give one of them up, we move between them. We cannot abide exculpation, because humans do truly terrible things. Think of the wet footprints on the floor. The pitchfork. The screwdriver next to the half-eaten sandwich. The stolen cheeseburgers. We cannot abide vengeance, because humans have terribly faulty bodies and endure terribly tragic circumstances. Think of the voices in the postpartum psychotic brain. The broken gene that robs mice and men of inhibition. The child abuse and prison rape. The eerily unbroken family legacy of murder. God vacillates in the Genesis story because we vacillate. Logic seems to demand an "or" but wisdom an "and."

When her mental state deteriorated, Andrea Yates's doctors had few treatments to offer her. In 2023, more than twenty years after Yates was psychologically undone by childbirth, the U.S. Food and Drug Administration finally approved the first oral medication specifically for postpartum depression, which affects approximately 10 percent of women who give birth. Zuranolone is a synthetic form of allopregnanolone, a hormone related to progesterone that affects both the body and the brain. The placenta produces allopregnanolone at high levels during pregnancy, bathing the developing fetus, and levels drop precipitously when a woman gives birth. The medication is not approved for the more serious condition of postpartum psychosis, which Yates also suffered from, but its development points to the possibility that one day medicine will have something more to offer women who, like her, are laid waste by pregnancy, childbirth, nursing.

Allopregnanolone has also been investigated in animals—guinea pigs and mice. If you give allopregnanolone to young male animals who are born prematurely, they are less likely to become hyperactive and aggressive. No such treatments are available for human animals, like Robert Harris, who are born early, born exposed to alcohol, born to parents with a history of severe antisocial behavior. There are, in fact, no medications that are approved, in any country, as treatments for conduct disorder, psychiatry's diagnostic label for children who persistently break rules and hurt others.

Is this pharmacological lack partly a consequence of how we, who are living in the long shadow of the Christian church, still typically frame the relationship between biology and behavior? If we, like Pelagius, insist that what is (im)moral can never be part of nature, then looking to science to discover new ways to treat the development of aggression and antisocial behavior in children would be nonsensical, a category error. Or if we, like

Augustine, insist that one's nature only makes one's suffering more necessary, then we have no motivation to look for treatment that might avert that suffering. But before Augustine, before Pelagius, the earliest Christians described moral salvation using the lexicon of medicine and proclaimed a Christ who healed the sick. A quarter century after my first introduction to the heretical sciences of psychology and genetics, this is the science I am now most interested in—science that can help heal.

What if we discover enough about the genetic causes of serious and persistent antisocial behavior that we can develop new pharmacological treatments to prevent it? What if, in the years to come, someone can write a paragraph about treatments that improve children's emotional empathy and impulse control as casually as I wrote a paragraph about antidepressant treatments at the beginning of this book? To treat her anxiety, the writer Lauren Slater "swallowed a pill designed through technology" and discovered herself "embedded in an animal world." This was the passage I quoted in the first chapter of this book. But everything about us is embedded in an animal world, including our most grievous sins and the punishments we have devised for them. When we refuse to look away from this potentially horrifying realization, we might begin to imagine new possibilities. One day, what could medicine offer children who might otherwise grow up to be like Jeffrey Landrigan or Robert Harris or Bradley Waldroup or Katharine Blake's cousin Scott or Adam Lanza or Andrea Yates?

Or Anders Breivik, who committed the worst mass murder in Norway's history?

As a young child, Anders Breivik exhibited textbook signs of callous-unemotionality and conduct disorder: He tortured animals, punched other children, deliberately urinated on neighbors' property. Neighborhood children were prohibited from playing with him, especially when their pets were around. His mother came to consider Breivik, the only child of a fleeting marriage, a "fundamentally evil and nasty child" as well as "clingy and difficult." "Aggressive, and nasty with it" read the case notes from a family counseling service who saw the family when Breivik was four years old. When Breivik grew up, he killed eight people with a bomb and then a few hours later hunted down and killed sixty-nine more, most of them teenagers, at an island summer camp.

Children who struggle with empathy and social connection show the most positive life outcomes when their parents are warm and consistent; Breivik's mother was the exact opposite. She had fled her abusive childhood home when she was a teenager, and she struggled with depression, anxiety, aggression, illogical thought patterns, and lability of mood for her entire life. Like her own mother, Breivik's grandmother, who was eventually hospitalized for paranoid schizophrenia, she became increasingly detached from reality as she aged. Social services recommended, given the instability of his home environment and his own psychological vulnerabilities, that Breivik be put into foster care, but this recommendation was never taken up. Instead, he got older and bigger, passing from a miserable and isolated childhood to an equally miserable and isolated adulthood.

After five years spent playing *World of Warcraft* in his mother's apartment, Breivik turned from a fantasy world ridden with orcs, taurens, and trolls to the fantasy world created by right-wing white nationalists, where the purity of the Nordic race was threatened by feminists, Muslims, and cultural Marxists. He began working on a manifesto, a declaration of war, which would make people see why punishment was necessary. Women were being raped by Muslim immigrants, he alleged. Wasn't rape a crime that called out for punishment? Weren't people advocating for multiculturalism complicit in this rape, traitors to their own society? People needed to know that you couldn't betray your country and get away with it. The most severe punishments, he wrote, should be reserved for class-A traitors: "political parties, trade unions, cultural institutions, and the media." Indifference to injustice was one of the worst sins. He would become the Punisher.

During Breivik's trial, psychological experts struggled to determine whether he was legally "sane" or "insane"—a simplistic psychiatric binary utterly outmatched by the complexity of his life. No one doubted that he suffered from an extreme abnormality in his ability to relate to other people, nor that his childhood circumstances had seriously hampered whatever hope of "normal" development he might have had. Yet he didn't fit neatly into any of the diagnostic boxes. Was he psychotic? Autistic? Narcissistic? Antisocial?

Testifying at trial, Ulrik Fredrik Malt, a professor of psychiatry, vividly commented on how poorly suited the diagnostic debate was for settling what people really wanted to know: "We have to ask ourselves as a society, as human beings and as psychologists. What is it that these [diagnostic] questions actually give us the answer to?" Coming up with the right diagnostic label, in his opinion, couldn't answer the more important question of how to treat Breivik as "a fellow human being." His membership in society—his fundamental humanness—was similarly emphasized by Åsne Seierstad, a Norwegian journalist who published an astonishing portrait of Breivik, his victims, and his trial. She titled her book *One of Us,* and that is, ultimately, what the court decided—that Breivik was a member of society, one of the collective "us," and so would be held accountable as such.

He was sentenced to an extra-high-security prison for twenty-one years, the maximum allowable sentence, with the option of additional time if he is then still deemed to be dangerous. He lives in three prison cells, one for exercising, one for studying, one for sleeping. He has access to video games and newspapers; he can play games with other inmates and talk on the phone. He can still vote. The prison operates under the "normality" principle, in which prisoners have the same rights and the same basic dignity in their living conditions as everyone else in society. To many outside Norway, the humanity and dignity with which Breivik is treated is shocking.

The Norwegian prison system was not always exceptional. Beginning in the 1960s, a prison abolition movement organized to abolish youth prison, forced prison labor, and long sentences for minor offenses; to minimize pre-trial detention; and to expand prisoners' access to media and correspondence. Life sentences were abolished in 1981. (Capital punishment had been abolished in 1902.) Subsequent reforms in 1998 and again in 2007 further focused on the reintegration of Norwegian prisoners through the improvement of prison conditions, psychotherapy, education, and job training. Norway now spends three times as much to incarcerate a prisoner as the United States, but it incarcerates people at one-tenth of the rate of the U.S. and one-third the rate of the U.K. Norwegian ex-prisoners are also one-third as likely to be arrested for a new crime within three years following release as prisoners from the U.K.—one of the lowest recidivism rates in the world.

In a cross-cultural study of how countries differ in their values, Norwegians were three times more likely than people in the United States—and seven times more likely than people in the U.K.—to say that severely punishing criminals is *not* an essential feature of democracy. Interestingly, Norway and the United

States do not markedly differ in their average beliefs about free will: Norwegians are just about as likely as Americans to say that they feel like they have "completely free choice and control over their lives" and that they can "decide their own destiny." Cherished beliefs (some might say illusions) about agency in one's own life do not have to be dashed, it seems, to soften the punitive drive.

Where Norway does differ markedly from the United States is belief in hell. Although Christianity became the dominant religion in Norway by the middle of the twelfth century, and most Norwegians remain nominally Christian, only 5 percent of the country reports regular religious attendance. And, crucially, less than 20 percent report they believe in hell. In contrast, more than 60 percent of Americans believe that some people will be punished for all eternity in a lake of fire, as do nearly 30 percent of people in the U.K., where belief in hell is highest among younger cohorts. Delight in torturing people in this life appears to go hand in hand with belief in a god who delights in torturing people in the next one.

Norway is not the only society that has had to deal with the aftermath of mass violence, and it is not the only society that has grappled with how to honor the seriousness of the wrongs done to the victims while still honoring the "one of us"–ness of the perpetrators. One of the most poignant, and tragically common, instances of this dilemma is how to deal with child soldiers returning from war. Just in the past few decades, hundreds of thousands of children, some as young as eight years old, have been used as combatants in wars around the world.

Ishmael Beah, for instance, was thirteen when a government army in Sierra Leone got him hooked on drugs and taught him how to shoot an automatic weapon. In his memoir, he reflected on how his terrible circumstantial luck did not, for him, cancel out his feelings of guilt and responsibility for the violence he perpetrated: "I was a kid when this happened. I wasn't psychologically developed enough to decide whether I would be a part of it or not, nor was there a choice in the situation. Nonetheless, I feel guilty about what I became and what I was forced to participate in or do or carry out."

The legal scholar Martha Minow, in her book *When Should Law Forgive?*, describes the dilemmas that child soldiers like Beah present in the aftermath of war and mass violence: "Laws usually do not have a category for individuals who are victims but come to engage in crimes or abuses of others." She suggests that "a new concept" needs to be devised "to acknowledge the complexity of individuals who have been both victims and perpetrators."

In the absence of cultural and legal concepts that acknowledge such complexity, many child soldiers have failed to be reintegrated into their communities. In Uganda, ethnographic researchers suggested that reintroductions by Western organizations that positioned abducted children solely as "innocent victims" failed to adequately address the "unavoidable change in the status of such children." The community, in response, "refuses to accept the idea that children are not accountable for the crimes that they have committed."

In contrast, studies of former child soldiers in Mozambique have identified better methods of reintegration into the community. Former soldiers participate in traditional cleansing ceremonies where they confess their violent acts—their wrongdoings must be exposed to the entire community—and then they seek to reconnect with protective spirits. These confessions, according to ethnographic reports, treat the returning children, at least for a time, as "wrongdoers rather than innocent victims." After the ceremonies are completed, however, cultural customs demand that people do not talk of the children's crimes again. Such ceremonies treat people who were severely victimized as children as accountable to the community, not for the sake of imposing additional suffering but with the goal of recognizing and restoring their relationships with others. As the philosopher Angela Smith summarized, "Being held responsible is as much a privilege as it is a burden. It signals that we are a full participant in the moral community."

No one imagines that children in war-torn countries had a choice about being abducted, addicted to drugs, forced to participate in violence. But the efforts to reintegrate former child soldiers into their communities pulls apart the usual terms of debates about free will, desert, and punishment. Being subject to forces beyond one's control does not mean innocence. Being held responsible by one's community for one's actions does not require the freedom to have done otherwise. But being held responsible does not require degradation, imprisonment, or excessive suffering. Rather, holding responsible is a social practice to address a change in people's relationships to one another and, possibly, repair them.

Of course, the dilemma posed by the dual identities of victim/perpetrator is not isolated to the theater of war, nor is that the only duality that troubles our rush to blame. All perpetrators were once children. All our constitutions were fashioned in childhood, by our environments and by our biology, with the boundary between biology and environment as porous and uncertain as the boundary between childhood and adulthood. The conjoined forces of nature and nurture shape behaviors commonly considered the province of the soul—addiction, violence, religion, punitiveness—and commonly modeled in nonhuman animals kept in laboratory cages.

Such duality undermines the logic of heaven and hell, which always presupposes a binary, which seeks to understand who deserves rescue and who deserves to suffer. Duality is horrifying. Of course it's horrifying; horror is the special sort of fear we feel at the dissolution of boundaries and the collapse of binaries. Extreme cases, like Yates, Breivik, and Beah, like Harris, Landrigan, and those with Brunner syndrome, horrify us not just because of the extremity of their acts but because the extremity of their life stories makes it impossible to forget their duality. They did horrible things, and they were at the mercy of forces beyond their control. Their choices were consequences of factors external to them, and they were held responsible. They remind us that sorting people based on the either/or logic of an afterlife that doesn't exist will always be an impossible task. As Minow put it: "Neither innocence nor guilt, all or nothing, fully offers the justice or the better future that individuals and societies need and deserve."

In April 2024, Travis and I find ourselves, once again, in rural Texas. This hundred-acre parcel of land used to be a cattle ranch; now it's owned by two intensive-care physicians who like to bowhunt. The only animals they keep are Lavender Wyandottes, chickens with a recessive gene that dilutes the black pigment in their feathers to a shimmering violet-gray. Local lore holds that a stony hill on their property, the highest place for miles around, was once an indigenous burial ground. Impossible to verify, but we are just a few miles from the Gault site, one of the oldest archeological sites in the Western Hemisphere. Humans have lived here, adjacent to the soils of the Blackland Prairie and the tool stones of the Edwards Plateau, for nearly twenty thousand years. We have come here, to this high place, to watch the moon blot out the sun.

I am pleased enough to be there but also irked to be missing a whole day of work. We were supposed to have an even larger group with us: Our friends flew from Vancouver to Houston, rented a car to drive to Austin, but then, discouraged by forecasts of cloud and rain, kept driving the next day, through east Texas and Arkansas and into Missouri. As a gay couple, they are, quite reasonably, nervous to venture into small unfamiliar towns, especially when hotel rooms are scarce. The Super 8 is rumored to be charging a thousand dollars a night along the path of totality. But they are committed to finding clear skies. It all strikes me as a bit excessive. "It's like you're on a pilgrimage!" I say in a tone that comes out as incredulous, but they look serenely back at me. "Yes, exactly. A pilgrimage. You'll see." I have seen a partial eclipse before, and I do not listen when they tell me that a total eclipse is a qualitatively different experience, that 100 percent is not just more than 95 percent, but other. Annie Dillard put it this way: "Seeing a partial eclipse bears the same relation to seeing a total eclipse as kissing a man does to marrying him."

I am skeptical. Travis and I decide to stick with the original plan, stick close to Austin. We are optimistic that we'll get lucky with the weather and that the clouds will part.

They do. The light, as the moon begins to occlude the sun, turns metallic. The landscape becomes a daguerreotype: silver-plated copper, iodized yellow-rose. Details are thrown into sharp relief. The plump leaves and smooth bark of the stunted persimmon trees are almost painful to behold. I remember that Moses's first glimpse of God was an incandescent wilderness shrub.

Have I ever seen plants so vividly before? Only on acid. My brain, a finely tuned prediction machine, is slowly being deprived of the one great constant on which all its predictive models have been trained—the sun in the sky. The resulting experience feels hallucinatory because that's what a hallucination is, a neural prediction that has become untethered from sensory input. "Whoa," says the sixteen-year-old boy in our group as he snacks on cheese puffs, Oreos, marshmallows. His younger brother scans the copperized earth for fossilized shells, remnants of when Texas was at the bottom of a vast inland sea.

The baby is drowsy in the hiking backpack. It is nap time, and we speculate when she will finally let herself fall asleep. Through our eclipse glasses we can see that the sun is almost entirely obscured. Without the glasses, though, we still cannot look directly at the sun; it is still too bright. We think that she will nod off when it finally gets dark.

But as the last sliver of sunlight disappears, she snaps fully awake. Her pacifier tumbles out of her open mouth, and she points upward. "Moon! Moooooon!" We, too, are agape.

Psychologists have defined awe as "an emotional response to perceptually vast stimuli that overwhelm current mental structures yet facilitate attempts at accommodation." In other words, it is too much; we do not understand. We try to understand, and still fail, but the trying changes us.

Awe has a particular physiological signature. Our sympathetic nervous systems slow; our muscles relax. Information processing also slows, becomes more careful and more detail-oriented. *Be still and know.* And our mouths open. The facial expression characteristic of awe is not limited to humans. There is a YouTube video of a chimpanzee liberated from long captivity. He sees the sky for the first time in decades, and his affective experience is instantly recognizable. He looks up, slack-jawed, delighted. The open mouth is characteristic of horror and its twin: wonder.

Scientific descriptions of awe characterize it as both a "self-diminishing" and a "self-expanding" emotion, an apparent contradiction that is telling: The experience of "I" depends on separation, boundaries, limits. In wonder, in awe, the imperious, rational, self-conscious ego, the one that is always insisting on its separateness, evaporates. You forget yourself. You forget to clench your jaw, to fix your face. As a result, the self that is unselfconscious, the experiencing self, can feel itself to be part of the ever-expanding world.

We stare at the black hole where the sun used to be. Time dilates. How long has it been? A second; an hour. I look over at the baby and realize my face is wet. I have, unbeknownst to myself, started crying.

"The stars aligned" is the idiom we English speakers use to express the mystery of why something happened—something wanted, something good, something out of our hands. The stars aligned: We are exposed to fortune on a grand scale. The stars aligned: This baby, this exact one, this precious one, the one yelling "Moon!," exists, and we are spinning on this rock together. The stars aligned: Our existence is dependent on impersonal forces that can sometimes be calculated but not controlled. The stars aligned: The universe is so much bigger than our notions of "desert" can contain. No one deserves to be born, yet here we are.

When I was still going to church, I could never quite manufacture a felt experience of God, but all the mystical feelings I could not conjure then now come unbidden. Smallness. Gratitude. Peace. Augustine thought it illogical that the Romans had temples to Fortuna, the goddess of luck, chance, fate, and life's cyclical nature, who was indifferent to humans' goodness or badness. "Why is she worshipped," he questioned, "who is thus blind, running at random on any one whatever, so that for the most part she passes by her worshippers, and cleaves to those who despise her?" But I feel more reverent here, experiencing one of nature's impersonal cycles, than in any church.

Fortuna was sometimes represented as both Jupiter's mother and Jupiter's child, a paradoxical duality I find everywhere. Identical twins, separated at birth, both believe their religious devotion is evidence of God's predestination of their natures. A docile bull's clone gores his owners. A rapist's sister helps discover the single DNA letter that has consigned him to a life of impulsivity and aggression. Some rats who have been shocked whenever they press the lever press it again and again. Some children who are spanked run toward those who hurt them and call it love. The retribution we imagine as divine is a pattern we can observe in insects, bacteria, immune cells. The forgiveness we long to give and receive is no less part of our evolved legacy than the sins that are forgiven. We are soul/bodies, who are born/made, who re/produce, who hold each other responsible for our luck/badness, as all the other agent/animals do.

The moon is covering the sun. One Hebrew word that is sometimes translated in the Bible as "to forgive" or "to absolve" is *kippēr*, which can mean "purge" or "wipe away" but can also mean "cover." Love covers a multitude of sins. The solar eclipse suggests to me a new sense in which something can be covered and, in so doing, a new understanding of the relationship between science and the forgiveness of wrongdoing.

Some thinkers write about science and blame as if they have only ever seen a partial eclipse. When the moon only partially obscures the sun, you still can't look at the sun with your naked eye. The sun is still blindingly bright, ferocious in its intensity. The nibbling away at its edges only increases one's appreciation of its light. In the same way, some discussions of scientific results end up reaffirming ordinary vengeance. We can glance at genetics—

and physics and neuroscience and psychology, too—and then go about our day, unchanged. Others, more rarely, write as if a total eclipse is permanent. To understand the science, they claim, is to obliterate blame, punishment, and responsibility entirely, to blot out the sun forever.

These descriptions fail to convey science's power and science's limits. What does biology have to do with blame? What does the moon have to do with the sun? Scientific results can eclipse blame—usually in part, and on rare occasions more completely. This obscuring is always temporary, always contingent on the observer's vantage point. But even if fleeting, even if particular to a certain perspective, the eclipse of blame can nonetheless be dumbfounding, disorienting, bewildering. It can leave us with an appreciation for the power of chance, awe at the vastness and interconnectedness of the world in which we are part, humility at the smallness of the self and the limits of human understanding. An eclipse is temporary, but we can be changed in more enduring ways by bearing witness to juxtaposition. We might find a more lasting reprieve from wrath in the aftereffects of wonder.

Puzzle

A coda, in which I endeavor to respond to the profound and irresolvable questions put to me by a man imprisoned for violence

When they say "The apple doesn't fall far from the tree," how do you interpret that statement?

APPLES ARE LIKE HUMANS: They don't breed true. If an apple falls from a Honeycrisp apple tree, and the seed germinates and grows into a tree, that tree will not necessarily produce Honeycrisp apples. It is impossible to predict what type of apple it will produce. Most of the time, the fruit will not look, or taste, anything like what is sold in the supermarket: About 1 in 80,000 apple trees grown from seed produce fruit that is as "good" as the parent's fruit.

Apples, like humans, don't breed true because of their genetics. Apples have two copies of each chromosome. Much of the DNA on one copy of each chromosome is different from the DNA on the other copy. Apple trees are unfruitful in isolation. They need to mix their genome with another tree to produce fruit. Each apple seed, then, contains a random draw from the two parental genotypes, and this genetic uniqueness results in fruit uniqueness. The rows of uniform apples that you see at the supermarket are possible only because growers do not allow apples to fall from the tree; that is, they do not grow apple trees from seed. Instead, they propagate apple trees by grafting. The trees that produce what we think of as apples are literally re-produced, asexually, as genetic clones of each other. Allowing apples to (re)produce sexually would lead to disaster for apple growers. The apple might not fall far from the tree, physically, but the apple contains within it the potential for a radically new fruit. I find it deliciously ironic that we comment on familial resemblance in people with reference to apples, another species in which offspring are different from their parents, not despite their genetic inheritance but because of it.

Fruits may fall close to the tree, but they are never meant to stay close to the tree. Fruit evolved as a mechanism for seed dispersal. Fruits attract animals, who eat them and spread their seeds to new territories. Apples are large and heavy. They evolved to attract ancient megafauna, who lived during the last Ice Age. Large animals can migrate long distances, introducing the apple seed to new landscapes far from the original tree. When environments are highly variable, genetic variability is an advantage: The odds of at least one apple seed surviving in a radically new environment are maximized when the seeds are all different from one another. What is an undesirable fruit from the perspective of today's commercial grower—tiny, lumpy, wizened, acerbic, inedible—might be the only seed that can survive in a novel environment. Genetic variety is necessary for adaptation, for evolution.

The other migrating animal that is tempted by apples is, of course, humans. Archeological sites suggest that humans have been collecting wild apples for more than ten thousand years. The modern apple brings together genetic material from several different populations of wild apples, which were transported along the Silk Road trade routes that connected East Asia with Europe. Even today, apples are exported across continents. The Golden Delicious apple, for instance, was a chance discovery, found on a volunteer tree sprouted from seed in West Virginia. This hedgerow accident can now be found far from the original tree: Millions of grafted Golden Delicious trees were planted in Europe in the aftermath of World War II, as part of the Marshall Plan to rebuild the European economy. The history of the apple, then, is simultaneously a history of humanity: desire, commerce, ingenuity, technology, aggression, cooperation. The history of the apple transcends the imaginary boundary between nature and culture.

The saying that "the apple doesn't fall far from the tree" has its own lineage. Poets and prophets and philosophers have long compared people to fruit-bearing plants. Jesus of Nazareth, in the book of Matthew, says:

> A good tree cannot bear bad fruit, nor can a bad tree bear good fruit. Every tree that does not bear good fruit is cut down and thrown into the fire. Therefore, by their fruits you shall know them.

People's fruits, in this metaphor, are their sins, their virtues, their behaviors. This passage expresses an essentialism about human character. Someone who is fundamentally wicked or separated from God's grace cannot help but manifest that inherent nature in their actions. Behavior is an expression of the soul, which will either be saved or condemned.

My people are from Texas, so I also grew up hearing about the proverbial apple that doesn't fall far from the proverbial tree. My dad's family had another, darker way of expressing a similar idea: "What's born in the blood can't be beat out the butt." Both sayings are used to comment on familial resemblance, to observe that like begets like. The sayings suggest not just recognition but resignation: "No wonder," and also "Nothing to be done." One judge wrote of a defendant convicted of a double homicide that knowing about the defendant's family history would have only served to convince the jury that he was a "bad apple from a bad tree and there was not hope for rehabilitation or redemption."

Used in this way, the apple proverb hybridizes essentialist ideas: the folk biology idea that humans breed true; the folk psychology idea that a person's values, capacities, and behaviors are essentially determined by their inherited biology; the Christian idea that humans are either figs or thistles, saved or damned. In this way, DNA is imagined to be both the mechanism for familial similarity and the replacement for the Christian soul (the "essence placeholder"), but even as the soul is jettisoned, the Christian binary that some people are bound for Heaven, others damned to Hell, is retained. As a scientist and as an atheist, I disagree with all three essentialist ideas. Humans do not breed true; our behavior is not essentially determined by our inherited biology; and our infinite variety cannot be categorized as saved or damned.

We can go back even further than Jesus to find plant-person comparisons. Pindar, an ancient Greek poet, wrote: "But human excellence / grows like a vine tree / fed by the green dew / raised up, among wise men and just, / to the liquid sky." In the lines directly preceding these, Pindar expresses his hope that he will praise what deserves praise and will "sow blame for wrongdoers." He clearly thinks that people can merit praise and blame by good or bad action, yet his comparison of human excellence to a vine deepens and complicates his moral judgments. Young vines are vulnerable. They need, for their growth, fair weather, nutritious soil, supporting trellises. Pindar calls attention to the luck of having received tender care and fortuitous circumstances.

The philosopher Martha Nussbaum, in her book *The Fragil-*

ity of Goodness, writes that Pindar's comparison of the good human life to a vine "confronts us with a deep dilemma in the poet's situation, which is also ours. . . . To what extent *can* we distinguish between what is up to the world and what is up to us, when assessing a human life?" She predicts that the dilemma posed by Pindar's ode will never stop being a dilemma, because humans will never stop being plantlike: "That I am an agent, but also a plant; that much that I did not make goes towards making me whatever I shall be praised or blamed for being . . . [this] I take to be not just the material of tragedy, but everyday facts of lived practical reason."

Pindar's poem suggests another interpretation of "the apple doesn't fall far from the tree." We could hear that line as a reminder of the fragility, struggle, and risk inherent in being a new human, growing from seed. A child is bequeathed a particular genome, lands in the shadow of her caregivers, must try to bloom where she is planted, and, in the end, is asked to prune herself. The lowly apple merits empathy and awe.

What makes a child go bad? Is it nature or nurture?

This question presumes that a child can be *bad*. Not just that their behavior contravenes some moral demand, but that whatever core thing a person *is* can be bad.

I don't think any child—*anyone*—is bad. Neither do I think everyone is essentially good. Rather, I think a person has no singular essence. By giving up faith in an immaterial soul, I have given up faith in a unitary "I." Perhaps here is where I have departed most radically from the Christianity of my youth: No one is either lamb or goat, wheat or tare, saved or damned. The complexity of any human life cannot be flattened to a single trait or attribute. I am not entirely defined by my race or ethnicity, or pattern of sexual attraction, or gender, or disability, or creed. Neither am I entirely defined by my genotype, nor by my behavior. Even my very worst behavior. And neither are you.

Allow me to rephrase your question: What makes a child behave badly—is it nature or nurture?

The trite answer, when psychologists are asked questions about nature or nurture, is that behavior is always shaped by both nature *and* nurture, and that indeed the business of separating the causes of an organism's behavior into those called "nature" versus those called "nurture" is nonsensical. DNA is an inert molecule, which depends on an environment for its causal powers and its very meaning. The genome is entangled in intricate developmental processes from the very beginning of life. It makes no sense, scientifically, to ask if a single person's behavior is due to nature or nurture, because nature and nurture are not separable.

This answer is technically true—and substantively evasive. Often, people's most pressing questions about behavior are not about the trait itself but about *differences* between people in their traits. Consider the mother of two sons, only one of whom has inherited the mutated "no-activity" copy of her *MAOA* gene and has committed acts of impulsive violence his entire life. The mother is not asking, "What is a complete scientific explanation of every developmental process that resulted in the aggression of this individual child?" but rather, "What is different about my child who is aggressive, compared to my child who is not?"

So let me rephrase your question again: What makes children different in how often they behave badly—is it nature or nurture?

Now we have arrived at a question that I can answer scientifically, although it's a different question than your original. And again, the answer is both nature and nurture. Environmental differences between children cause differences in their behavior. Children who are exposed to more lead, for instance, are more likely to behave aggressively, an environmental effect that operates through the toxin's effects on brain development. Genetic differences between children also cause differences in their behavior. Some of these genetic differences are very rare mutations with large effects, as in Brunner syndrome. Most genetic differences have very small effects that accrue like grains of sand. Many genetic effects operate through, and are moderated by, the social environment. Parents respond more harshly to daring and brash children, for instance, and children who experience harsh parenting become more antisocial. Children who are at genetic risk for antisocial behavior show the best outcomes when their adoptive parents are warm and nonpunitive.

But is this what you really wanted to know? Why do questions about nature or nurture seem so pressing? From some vantage points, the question might appear to be merely of academic interest. If one is a committed determinist and free-will skeptic, the crucial observation is that behavior is caused; whether the causes are conventionally thought of as nature or nurture makes not one whit of difference. Yet this deterministic logic has not dispelled interest in the nature-nurture debate.

One reason the nature-nurture debate seems so pressing is that it appears to bear on questions of change. Is the question behind your question, "Can I change?" If differences are "genetic," they are commonly held to be fixed, immutable, unchangeable. The philosopher Evelyn Fox Keller proposed: "Perhaps we should rephrase the nature-nurture question, and ask, instead, how malleable is a given trait, at a specified developmental age?" Empirically, there is no simple correspondence between the question of whether a phenotype is affected by genetic differences and whether that phenotype is plastic, mutable, adapting.

For example: We treat inborn disorders of metabolism with specialized diets at infancy; we treat highly heritable myopia with eyeglasses; we give growth hormone to children whose genes have caused them to have very short stature. On the other hand, we have no cures for blindness caused by exposure to poison sumac; many people whose anxiety and depression are caused by trauma struggle to find relief. Indeed, one of the major goals for medical genetics is to find "druggable" targets: Scientists want to identify genetic contributions to differences in disease risk precisely so they can change things.

But the intuition that genetically caused behavior is less changeable persists. This conflation of genetics with incorrigibility is yet another legacy of the doctrine of original sin, which implicates the rebellious body, the fallen flesh, as the primary barrier to human virtue. We would do better, we would *will* better, Augustine said, if we hadn't been punished by God with a body that betrays us. And even now, we imagine the flesh—refashioned as the gene—to be an insurmountable obstacle to human perfectibility. What would it take to recast the flesh, the gene, as something other than the enemy of heaven, to see it as the mechanism for plasticity, adaptation, evolution? To associate "nature" not with fixity, stuckness, impotence? To achieve what Louise Glück called "substitution / of the immutable / for the shifting, the evolving"? We would have to revise ancient habits of mind.

For the early Christian church, the question of nature versus nurture was a question of saint-or-heretic, life-or-death debate. Augustine's revolutionary doctrine that sin could be physically inherited helped vindicate belief in a good Creator but cast humans as inexorably flawed. Objections to Augustine's theology insisted on humanity's potential for transformation and reinvention by drawing a strict boundary between nature and moral behavior. We inherited the idea that the flesh is the enemy of moral transformation from millennia of Christian thought.

I thought, when I started writing this book, that the science linking genetic difference with behavioral difference was disturbing because it faintly echoed Christian ideas. I now see that it goes much deeper than that, that the question of whether behavior is influenced by nature dredges up all the paradoxes that Christianity was trying to solve. Why do we suffer? Why is there evil? How can we maintain faith in goodness in a cruel, capricious world? Can we have hope for the future?

Now when you ask, "What makes a child go bad, is it nature or nurture?," I wonder if you're really asking me, "Was it all my fault? Am I all bad? Do I have hope for the future?" That is not a scientific question to which I can provide a scientific answer. But yes, I believe, by faith and not by sight, that you do have hope. I believe everyone does. Not, I should stress, hope that necessarily takes the form of a single quasi-magical conversion experience that permanently transforms your will. Odysseus's hope against the enchantments of the Sirens was to be lashed to the mast, restrained by his community in acknowledgment of the limits of his will. The hope for the family with Brunner syndrome will come from science, not prayer—unless you consider science, as a method of disciplined attention and openness to mystery, a form of prayer. Hope can take many forms.

I have spent much of this book listing my dissatisfactions and disagreements with the Christian doctrine of my youth. I don't think that our true self is an immaterial soul that is either saved or damned. I don't think sex is the source of our corruption. I don't think babies deserve to suffer. I don't think our only hope is in Christian salvation. Yet, in the end, the line that rings in my head is from the Gospel. Maybe that speaks to the hold that religion can have on you, long after you think you've given it up. Or maybe it speaks to the power of story, of words. Words of mercy rather than blame; of hope rather than despair; of evolution rather than perdition. Words we will not hear from a god and long to hear from each other:

Neither do I condemn you. Go and sin no more.

Why would a young boy of sixteen attack a total stranger, a female, at knife point in broad daylight at a busy intersection and make her drive against her will, to sexually assault her? What would drive a boy to do such a thing?

I cannot answer that question. No one can, not even you. Humans are storytellers, but the stories we tell about why we do what we do are just that—stories. In *Between Vengeance and Forgiveness*, the legal scholar Martha Minow describes the truth commissions and war-crimes trials that occurred in the aftermath of mass violence around the world. She writes: "There is in these stories a lack of closure, and the impossibility of balance and satisfaction, in the face of incomprehensible human violence." Human violence does not have to occur on a mass scale to be incomprehensible; human behavior does not have to be violent to resist closure, satisfaction, certainty. There is no answer to the question of "why" that will undo the past.

And we have again returned to a crucial distinction, between understanding the development of a specific person's specific behavior at a specific moment in time, and understanding what things, on average, cause differences in behavior between people in the long run. Psychology, sociology, economics are almost exclusively concerned with the latter. I can tell you a slew of variables that, on average, are related to differences in the probability of violent behavior, including (in no particular order and this list is not comprehensive and these variables are overlapping and intersecting): male sex, adolescent age, exposure to lead and other neurotoxins, living in a neighborhood with concentrated poverty and high crime rates, head injury, being a victim of violence, antisocial peers, problems with executive function, learning disability, childhood ADHD, dark triad personality traits, fearlessness, negative emotionality, access to weapons, substance use, and having a high number of risk-increasing genetic variants.

Which of those things characterizes you? What has been important for shaping your behavior that is not on that list? Which of these things would have had to be different for you not to have kidnapped, assaulted, traumatized that girl? In a world of probabilities, the leap from the average to the individual, from science to biography, is always difficult and often impossible.

The girl. She is the person whose story is missing here. She must also have asked: Why is there evil? Why did I suffer like this? How do I live in this cruel, capricious world? You are asking, "Why did I do this?," which seems necessary, and also necessarily incomplete. I hope you are also asking, "What now is required of me?" I hope you are also asking: "What would she have me do?" Perhaps, as other victims have described, she would have you live your life in service to other people.

There is an obvious, inescapable tension to ending here. But I think you might already know this. On the last page of your letter, you included a full-page photocopy of a trompe-l'oeil pencil drawing by the artist J. D. Hillberry, titled "Putting It Together." The drawing is of an incomplete jigsaw puzzle. The connected pieces begin to form a portrait of a somber-looking young child with downcast eyes. The illusion of three-dimensional space is isolated to just one part of the picture: The right hand of the child reaches out of the surface, holding a piece of the puzzle, assembling himself.

The drawing expresses the impossible paradox of being human. How can a constituted being constitute their own being? We are animals who treat each other as agents—animals, both predatory and ruminant, who aspired to be like gods. We have eaten the fruit of the tree of knowledge of good and evil, and our eyes of self-consciousness are open. We are the object and the subject of the puzzle.

Most pieces of the puzzle, though, remain missing.

Acknowledgments

THANK YOU TO ELIZABETH Barnes, Suzette Ermler, Pamela Hieronymi, Joshua Knobe, Brian Leiter, Michael McKenna, Alfred Mele, Erik Parens, Rebecca Saxe, and Eric Turkheimer for helpful discussions. I am grateful for the opportunity to collaborate on scientific studies of addiction and antisocial behavior with members of the Externalizing Consortium, and for grant support from the National Institutes of Health and the Jacobs Foundation. I am also grateful to the members of the Developmental Behavior Genetics lab, including Maggie Clapp, Diego Londoño-Correa, Abby deSteiguer, Yavor Dragostinov, Garrett Ennis, Rachel Hsu, Katarina Jakimier, and Sofia Semyrenko, for feedback on earlier drafts of parts of this manuscript (and for keeping things going while I was very distracted). Thank you to my agents, Mel Flashman and Will Francis, for believing in this book project, and to my editors, Marie Pantojan and Jenny Lord, for your patience and incisiveness.

Thank you to Sara Beckmann, Jen Doleac, Ben Domingue, Paul Eastwick, Sam Gosling, Gideon Lewis-Kraus, Natalia

Wulfe, and members of the Fictitious Friends Book Coven for your advice, recommendations, encouragement, and diversions. Thank you to my children—Jonah, Rowan, and Lavinia—for your vibrancy, humor, wisdom, and love. And thank you to Travis, my beloved, for all the ways you care.

Notes

ix **"Let the reader"**: St. Augustine, *The Trinity*, as quoted in Garry Wills, *Saint Augustine* (New York: Viking, 1999), xiv.
ix **"Without contraries"**: William Blake, *The Marriage of Heaven and Hell*, ed. Will Jonson (CreateSpace Independent Publishing Platform, 2014).

Desert

4 **Michael Pollan's book on LSD:** Michael Pollan, *How to Change Your Mind: What the New Science of Psychedelics Teaches Us About Consciousness, Dying, Addiction, Depression, and Transcendence* (New York: Penguin Press, 2018), 14. Pollan attributed to Carl Jung the idea that "it is not the young but people in middle age who need to have an 'experience of the numinous' to help them negotiate the second half of their lives."
5 **"We have come to think, lately":** Lauren Slater, "Black Swans," *The Missouri Review* 19, no. 1 (1996): 29–46, doi.org/10.1353/mis.1996.0015.
5 **serotonin system's effects on neural plasticity:** Rachael Sumner and Kacper Lukasiewicz, "Psychedelics and Neural Plasticity," *BMC Neuroscience* 24, no. 1 (June 30, 2023): 35, doi.org/10.1186/s12868-023-00809-0; Igor Branchi, "The Double Edged Sword of Neural Plasticity: Increasing Serotonin Levels Leads to Both Greater Vulnerability to Depression and Improved Capacity to Recover," *Psychoneuroendocrinology* 36, no. 3 (April 2011): 339–51, doi.org/10.1016/j.psyneuen.2010.08.011.
7 **Marcia Powell died:** *No Human Involved*, 2017, vimeo.com/ondemand/nohumaninvolved.
7 **She was found unresponsive:** Mark A. Fischione, *Report of Autopsy*, no. 09-02884, Office of the Medical Examiner, August 19, 2009, media.phoenixnewtimes.com/3804678.0.pdf; Stephen Lemons, "Marcia Powell's Autopsy Report Rules Cage Death in ADC Custody an 'Accident,'" *Phoenix*

New Times, August 31, 2009, phoenixnewtimes.com/news/marcia-powells-autopsy-report-rules-cage-death-in-adc-custody-an-accident-6501501.

7 **Three guards:** James Ridgeway, "Dog Days Turn Deadly in America's Prisons," *Mother Jones,* September 1, 2009, motherjones.com/criminal-justice/2009/09/dog-days-turn-deadly-americas-prisons.

8 **No one in the Arizona:** Sixteen employees were fired, suspended, or demoted. See Associated Press, "Disciplinary Action Taken After Inmate Marcia Powell Dies from Being Put in Outdoor Cell in Heat," *New York Daily News,* September 23, 2009, nydailynews.com/2009/09/23/disciplinary-action-taken-after-inmate-marcia-powell-dies-from-being-put-in-outdoor-cell-in-heat.

8 **"an expanse of deep and arid sand":** Dante, *The Inferno,* trans. Robert Hollander and Jean Hollander (New York: Anchor, 2002), XIV.13, 259.

10 **"that class of the terrifying":** Sigmund Freud, "The 'Uncanny,'" *Collected Papers,* vol. 4, trans. Alix Strachey (London: Hogarth Press, 1925), 369–70.

11 **"There's nothing you can do about it":** This line is said by the character Wintergreen in Joseph Heller's novel *Closing Time: The Sequel to Catch-22* (Riverside: Simon & Schuster, 1995). I learned of Wintergreen in an essay by the philosopher Galen Strawson that surveyed common philosophical positions on questions of determinism, free will, and moral responsibility. Strawson explained that "a strange minority" of philosophers assert that "we can be morally responsible for what we do even though we are not free agents. This view . . . has a kind of existential panache, and appears to be embraced by Wintergreen . . . , as well as by some Protestants." Strawson's passing reference to Protestants was formative for my thinking. Galen Strawson, "Luck Swallows Everything," *Times Literary Supplement,* June 26, 1998, also available at naturalism.org/philosophy/free-will/luck-swallows-everything.

11 **"That thought—'the miscreant'":** Friedrich Nietzsche, *On the Genealogy of Morals,* ed. Robert C. Holub, trans. Michael A. Scarpitti (London: Penguin Classics, 2014).

11 **before we invented villains:** Friedrich Nietzsche, *Thus Spake Zarathustra: A Book for All and None* (Macmillan and Company, 1896). "'Enemy' you shall say, but not 'villain'; 'sick' you shall say, but not 'scoundrel'; 'fool' you shall say, but not 'sinner.'" I am grateful to an essay by Dr. Brian Leiter titled "Blame and Christianity" for drawing my attention to this passage. See Brian Leiter, *Boston Review,* July 10, 2013, bostonreview.net/forum_response/brian-leiter-blame-and-christianity.

11 **"Suppose you sin":** Leviticus 5:17, New Living Translation.

12 **"This punishment that I have laid":** Sophocles, *The Oedipus Cycle: An English Version* (New York: Harcourt, Brace, 1949), archive.org/details/oedipuscycle0000unse.

Sin

13 **one of the oldest prisons in Texas:** Robert Perkinson, *Texas Tough: The Rise of America's Prison Empire* (New York: Picador, 2010).

Notes

13 **"God-forsaken hole":** "The Recent Eastham Prison Breaks," *Sheriffs' Association of Texas Official Magazine* 6, no. 8 (August 1937): 18–19, cited in Perkinson, *Texas Tough.*

14 **It was from an article:** Jennifer Latson, "Are Some People Born Lucky? A UT Psychologist Argues Inequality's Genetic Roots," *Texas Monthly,* September 2021, texasmonthly.com/news-politics/genetic-lottery-kathryn-paige-harden.

15 **my colleagues and I published a paper:** Richard Karlsson Linnér et al., "Multivariate Analysis of 1.5 Million People Identifies Genetic Associations with Traits Related to Self-Regulation and Addiction," *Nature Neuroscience* 24, no. 10 (October 2021): 1367–76, doi.org/10.1038/s41593-021-00908-3.

16 **Philip K. Dick novella:** Philip K. Dick, *The Minority Report* (New York: Pantheon, 2002).

17 **"the vice of lechery":** Dante, *The Inferno,* V.55, 93.

17 **"those who by violence":** Dante, *The Inferno,* XII.48, 221.

17 ***Diagnostic and Statistical Manual:*** American Psychiatric Association, *Diagnostic and Statistical Manual of Mental Disorders, Text Revision DSM-5-TR,* 5th ed. (Washington, D.C.: American Psychiatric Publications, 2022).

18 **"Because he aspired":** Dante, *The Inferno,* XX.38–39, 363.

19 **"the innate sinful depravity":** Jonathan Edwards, *Original Sin,* The Works of Jonathan Edwards Series, vol. 3, ed. Clyde A. Holbrook (New Haven: Yale University Press, 1970).

19 **"the human propensity":** Francis Spufford, *Unapologetic: Why, Despite Everything, Christianity Can Still Make Surprising Emotional Sense* (New York: HarperOne, 2013), 27.

20 **"The contradiction of the moral":** Paul Tillich, *A History of Christian Thought* (New York: Touchstone, 1972), 124.

20 **"What is natural":** St. Augustine, *Opus Imperfectum Contra Julianum,* as cited in Elaine Pagels, *Adam, Eve, and the Serpent: Sex and Politics in Early Christianity* (New York: Quality Paperback Book Club, 2005), 7, 135, 144.

20 **"The transmission of hereditary":** Henry Chadwick, *The Early Church,* The Pelican History of the Church (London: Hodder and Stoughton, 1968), 232.

21 **You are to blame:** This sentence is from Nietzsche: "Everyone who suffers, in fact, seeks instinctively the cause of his suffering; to put it more precisely, an agent—to put it more precisely still, a *culprit,* some guilty party susceptible to pain—in brief, some living thing upon which either actually or in effigy, he can, on any pretext, vent his emotions. . . . 'I suffer: someone is to blame'—all sick sheep think this. But his shepherd, the ascetic priest, says to him, 'Quite so, my sheep, it must be the fault of someone; but, you yourself are that someone, you alone are to blame—*you yourself are to blame for yourself.*" Nietzsche, *On the Genealogy of Morals,* 113–14.

22 **"I do not do":** Romans 7:19 (New International Version).

22 **"metaphysics of the hangman":** Friedrich Nietzsche, *Twilight of the Idols,* trans. Richard Polt (Indianapolis: Hackett, 1997), 35.

23 **People who believe:** Anja Hilbert, "Weight Stigma Reduction and Genetic Determinism," *PLoS ONE* 11, no. 9 (September 15, 2016): e0162993, doi.org/10.1371/journal.pone.0162993; Joanne A. Rathbone et al., "How Conceptualizing Obesity as a Disease Affects Beliefs About Weight, and Associated Weight Stigma and Clinical Decision-Making in Health Care," *British Journal of Health Psychology* 28, no. 2 (2023): 291–305, doi.org/10.1111/bjhp.12625.

23 **"still murky gene-environment interactions":** Julia Belluz, "Scientists Don't Agree on What Causes Obesity, but They Know What Doesn't," *The New York Times*, November 21, 2022, nytimes.com/2022/11/21/opinion/obesity-cause.html.

23 **"help us see that metabolism":** Jia Tolentino, "Will the Ozempic Era Change How We Think About Being Fat and Being Thin?," *The New Yorker*, March 20, 2023, newyorker.com/magazine/2023/03/27/will-the-ozempic-era-change-how-we-think-about-being-fat-and-being-thin.

24 **"Chemical Imbalance":** *Original Zoloft Commercial*, 2009, youtube.com/watch?v=twhvtzd6gXA.

24 **believe "biogenetic" explanations:** Nick Haslam and Erlend P. Kvaale, "Biogenetic Explanations of Mental Disorder: The Mixed-Blessings Model," *Current Directions in Psychological Science* 24, no. 5 (October 1, 2015): 399–404, doi.org/10.1177/0963721415588082.

24 **autism was widely blamed on parents:** Mitzi M. Waltz, "Mothers and Autism: The Evolution of a Discourse of Blame," *AMA Journal of Ethics* 17, no. 4 (April 2015): 353–58, doi.org/doi:10.1001/journalofethics.2015.17.4.mhst1-1504.

24 **"just happening to defrost":** Patty Douglas, "Refrigerator Mothers," *Journal of the Motherhood Initiative for Research and Community Involvement*, July 31, 2014, jarm.journals.yorku.ca/index.php/jarm/article/view/39328.

24 **"Strong evidence against":** David G. Amaral, "Examining the Causes of Autism," *Cerebrum* (January 1, 2017): cer-01-17.

24 **"Just because your child":** Danielle Dick, *The Child Code: Understanding Your Child's Unique Nature for Happier, More Effective Parenting* (New York: Avery, 2021).

25 **poverty is an inevitable consequence:** Matthew Desmond, *Poverty, by America* (New York: Crown, 2023); Richard J. Herrnstein and Charles Murray, *The Bell Curve: Intelligence and Class Structure in American Life* (New York: Free Press, 1996). I discuss the uses and misuses of genetics in relation to debates about poverty and social inequality in my first book. See K. Paige Harden, *The Genetic Lottery: Why DNA Matters for Social Equality* (Princeton: Princeton University Press, 2021).

25 **"When [scientific] inquiry":** Joe Dallas, "Born Gay?," *Christianity Today*, June 22, 1992, 20.

25 **shifting attitudes about gay rights:** Elizabeth Suhay and Toby Epstein Jayaratne, "Does Biology Justify Ideology? The Politics of Genetic Attribution," *Public Opinion Quarterly* 77, no. 2 (2013): 497–521, doi.org/10.1093/poq/nfs049.

26 **One highly publicized legal case:** Jan Hoffman, "She Went to Jail for a Drug Relapse. Tough Love or Too Harsh?," *The New York Times,* June 4, 2018, nytimes.com/2018/06/04/health/drug-addict-relapse-opioids.html; "*Commonwealth v. Eldred,* 101 N.E.3d 911 (Mass. 2018)," *Harvard Law Review* 132, no. 7 (May 2019): 2074–81; *Commonwealth v. Eldred,* 101 N.E. 3d 911 (Mass. 2018).

26 **central question of the case:** "If Addiction Is a Disease, Why Is Relapsing a Crime?," *The New York Times,* May 30, 2018, nytimes.com/2018/05/29/opinion/addiction-relapse-prosecutions.html.

26 **"there is no necessary connection":** Gene M. Heyman et al., "Brief of Amici Curae of 11 Addiction Experts in Support of Appellee," ma-appellatecourts.org/pdf/SJC-12279/SJC-12279_08_Amicus_Morse_Brief.pdf. One of the co-authors of this brief, Dr. Sally Satel, spent nearly twenty years publicly advocating against restrictions of opioid drug prescribing and was later found to have ties with Purdue Pharma, whose media strategy aimed to portray people who became addicted to opioid drugs—not the drugs themselves—as the problem fueling the opioid epidemic. See David Armstrong, "Inside Purdue Pharma's Media Playbook: How It Planted the Opioid 'Anti-Story,'" *ProPublica,* November 19, 2019, propublica.org/article/inside-purdue-pharma-media-playbook-how-it-planted-the-opioid-anti-story.

26 **"come tell your mama goodbye":** "Can Your Genes Make You Murder?," *Morning Edition,* NPR, July 1, 2010, npr.org/2010/07/01/128043329/can-your-genes-make-you-murder.

27 **"A person doesn't choose":** "Can Your Genes Make You Murder?," *Morning Edition.*

27 **Avshalom Caspi and his colleagues:** Avshalom Caspi et al., "Role of Genotype in the Cycle of Violence in Maltreated Children," *Science* 297, no. 5582 (August 2, 2002): 851–54, doi.org/10.1126/science.1072290.

27 **in the pages of academic journals:** John K. Hewitt, "Editorial Policy on Candidate Gene Association and Candidate Gene-by-Environment Interaction Studies of Complex Traits," *Behavior Genetics* 42, no. 1 (January 2012): 1–2, doi.org/10.1007/s10519-011-9504-z.

27 **in the courtroom:** Nita A. Farahany, Roderick T. Kennedy, and Brandon L. Garrett, "Genetic Evidence, MAOA, and *State v. Yepez,*" *New Mexico Law Review* 50, no. 3 (2020): 469–87.

27 **The study did not include:** Laramie E. Duncan and Matthew C. Keller, "A Critical Review of the First 10 Years of Candidate Gene-by-Environment Interaction Research in Psychiatry," *American Journal of Psychiatry* 168, no. 10 (October 1, 2011): 1041–49, doi.org/10.1176/appi.ajp.2011.11020191.

28 **less than a percentage point:** Richard Karlsson Linnér et al., "Multivariate Analysis of 1.5 Million People Identifies Genetic Associations with Traits Related to Self-Regulation and Addiction," *Nature Neuroscience* 24 (2021): 1367–76, doi.org/10.1038/s41593-021-00908-3; Jorim J. Tielbeek et al., "Uncovering the Genetic Architecture of Broad Antisocial Behavior Through a Genome-Wide Association Study Meta-Analysis," *Molecular*

Psychiatry 27, no. 11 (November 2022): 4453–63, doi.org/10.1038/s41380-022-01793-3.

28 **very small effects**: Christopher F. Chabris et al., "The Fourth Law of Behavior Genetics," *Current Directions in Psychological Science* 24, no. 4 (August 1, 2015): 304–12, doi.org/10.1177/0963721415580430.

28 **Prosecutors were "flabbergasted"**: "Can Your Genes Make You Murder?," *Morning Edition*.

29 **rarely worked out in the defendant's favor**: Deborah W. Denno, "Courts' Increasing Consideration of Behavioral Genetics Evidence in Criminal Cases: Results of a Longitudinal Study," *Michigan State Law Review* 2011, no. 3 (2011): 967–1050; Deborah W. Denno, "What Real-World Criminal Cases Tell Us About Genetics Evidence," SSRN Scholarly Paper (Rochester: Social Science Research Network, August 12, 2013), papers.ssrn.com/abstract=2308890; Sally McSwiggan, Bernice Elger, and Paul S. Appelbaum, "The Forensic Use of Behavioral Genetics in Criminal Proceedings: Case of the MAOA-L Genotype," *International Journal of Law and Psychiatry* 50 (January 1, 2017): 17–23, doi.org/10.1016/j.ijlp.2016.09.005.

29 **"evidence of mere genetic susceptibility"**: *State v. Yepez*, 2021 NMSC 10 (NM 2021).

29 **"In high-stakes criminal trials"**: Nicholas Scurich and Paul S. Appelbaum, "*State v. Yepez*: Admissibility and Relevance of Behavioral Genetic Evidence in a Criminal Trial," *Psychiatric Services* 72, no. 7 (July 2021): 853–55, doi.org/10.1176/appi.ps.202100226.

29 **people are *more* punitive**: Meredith Meyer et al., "Genetic Essentialist Beliefs About Criminality Predict Harshness of Recommended Punishment," *Journal of Experimental Psychology: General* 151, no. 12 (December 2022): 3230–48, doi.org/10.1037/xge0001240.

31 **"I felt certainty diminish"**: Katharine Blake, *The Uninnocent: Notes on Violence and Mercy* (New York: Farrar, Straus and Giroux, 2021), 163, 187.

31 **"We need to accept"**: Robert M. Sapolsky, *Determined: A Science of Life Without Free Will* (New York: Penguin Press, 2023), 403.

32 **"no reason . . . not to continue"**: Kevin J. Mitchell, *Free Agents: How Evolution Gave Us Free Will* (Princeton: Princeton University Press, 2023), 283.

34 **"Suffer the little children"**: Matthew 19:14 (New International Version).

Luck

37 **"You can't stop people"**: Martin Amis, *London Fields* (New York: Vintage, 1991).

38 **"What is it like to be a bat?"**: Thomas Nagel, "What Is It Like to Be a Bat?," *Philosophical Review* 83, no. 4 (October 1974): 435, doi.org/10.2307/2183914.

40 **"mark of Cain"**: Adrian Raine, *The Anatomy of Violence: The Biological Roots of Crime* (New York: Vintage, 2014).

41 **unseen anomalies:** Tímea Csulak et al., "Increased Prevalence of Minor Physical Anomalies Among the Healthy First-Degree Relatives of Bipolar I Patients—Results with the Méhes Scale," *Frontiers in Psychiatry* 12 (April 29, 2021): 672241, doi.org/10.3389/fpsyt.2021.672241.

41 **when they were preschoolers:** Mary F. Waldrop and Jacob D. Goering, "Hyperactivity and Minor Physical Anomalies in Elementary School Children," *American Journal of Orthopsychiatry* 41, no. 4 (1971): 602–7, doi.org/10.1111/j.1939-0025.1971.tb03219.x.

41 **manifestations of aggression change:** Isaac T. Petersen, Daniel Ewon Choe, and Brandon LeBeau, "Studying a Moving Target in Development: The Challenge and Opportunity of Heterotypic Continuity," *Developmental Review* 58 (December 2020): 100935, doi.org/10.1016/j.dr.2020.100935.

41 **Around the world and across ages:** Louise Arseneault et al., "Minor Physical Anomalies and Family Adversity as Risk Factors for Violent Delinquency in Adolescence," *American Journal of Psychiatry* 157, no. 6 (June 2000): 917–23, doi.org/10.1176/appi.ajp.157.6.917; Sarnoff A. Mednick and Elizabeth S. Kandel, "Congenital Determinants of Violence," *Bulletin of the American Academy of Psychiatry & the Law* 16, no. 2 (1988): 101–9; Daniel Pine et al., "Minor Physical Anomalies: Modifiers of Environmental Risk for Psychopathology," *Publications and Research*, January 1, 1997, academicworks.cuny.edu/cc_pubs/293.

41 **the "Stocking Strangler":** Raine, *The Anatomy of Violence*.

42 **If we *Expect Better*:** Emily Oster, *Expecting Better: Why the Conventional Pregnancy Wisdom Is Wrong—and What You Really Need to Know* (New York: Penguin, 2013); Ruth Anne Hammond, *Respecting Babies: A New Look at Magda Gerber's RIE Approach* (Washington, D.C.: Zero to Three, 2009).

42 **the bookstore offers you:** Hunter Clarke-Fields, *Raising Good Humans: A Mindful Guide to Breaking the Cycle of Reactive Parenting and Raising Kind, Confident Kids* (Oakland: New Harbinger Publications, 2019); Sonora Jha, *How to Raise a Feminist Son: Motherhood, Masculinity, and the Making of My Family* (Seattle: Sasquatch Books, 2021); Joe Newman, *Raising Lions: The Art of Compassionate Discipline* (self-published, 2010); Melinda Wenner Moyer, *How to Raise Kids Who Aren't Assholes: Science-Based Strategies for Better Parenting—from Tots to Teens* (New York: Putnam, 2021).

43 **Aristotle introduced the word *catharsis*:** "Catharsis," Encyclopaedia Britannica, March 13, 2025, britannica.com/art/catharsis-criticism.

43 **Freud later adapted the term:** Jean-Michel Vives, "Catharsis: Psychoanalysis and the Theatre," *International Journal of Psycho-Analysis* 92, no. 4 (August 2011): 1009–27, doi.org/10.1111/j.1745-8315.2011.00409.x.

44 **Lionel Shriver's novel:** Lionel Shriver, *We Need to Talk About Kevin* (New York: Harper Perennial, 2011).

44 **Solomon interviews Sue and Tom Klebold:** Andrew Solomon, *Far from the Tree: Parents, Children, and the Search for Identity* (New York: Scribner, 2013).

45 "stark fiction": Bernard Williams, "The Women of Trachis: Fictions, Pessimism, Ethics," in *The Sense of the Past: Essays in the History of Philosophy*, ed. Robert B. Louden and Paul Schollmeier (Princeton: Princeton University Press, 2009), 49–59, doi.org/10.1515/9781400827107.49.

47 "affected" males in this family: H. G. Brunner et al., "Abnormal Behavior Associated with a Point Mutation in the Structural Gene for Monoamine Oxidase A," *Science* 262, no. 5133 (October 22, 1993): 578–80, doi.org/10.1126/science.8211186.

47 can have extremely large effects: Lucia Corte et al., "Trumpet Plots: Visualizing the Relationship Between Allele Frequency and Effect Size in Genetic Association Studies," *GigaByte* 2023 (September 1, 2023): gigabyte89, doi.org/10.46471/gigabyte.89; Ju-Hyun Park et al., "Distribution of Allele Frequencies and Effect Sizes and Their Interrelationships for Common Genetic Susceptibility Variants," *Proceedings of the National Academy of Sciences* 108, no. 44 (November 2011): 18026–31, doi.org/10.1073/pnas.1114759108.

47 "There is no such thing": Solomon, *Far from the Tree*.

48 The *MAOA* gene: The scientific convention is that human gene names are italicized and written in uppercase letters (*MAOA*); mouse and rat gene names are italicized and only the first letter is upper case (*Maoa*), and the proteins these genes code for are set in roman and written in all uppercase (MAOA). "Gene and Protein Nomenclature," *Molecular Human Reproduction*, accessed March 7, 2025, academic.oup.com/molehr/pages/Gene_And_Protein_Nomenclature.

49 explosive outbursts of anger: Brunner et al., "Abnormal Behavior Associated with a Point Mutation in the Structural Gene for Monoamine Oxidase A."

49 In Portugal, a man is arrested: A. L. Falcão et al., "Brunner Syndrome: From Genetics to Psychiatry—Case Report," *European Psychiatry* 66, no. S1 (March 2023): S886, doi.org/10.1192/j.eurpsy.2023.1877.

49 In France, a seven-year-old boy: Amélie Piton et al., "A Second Mutation in MAOA Identified by Targeted High-Throughput Sequencing in a Family with Altered Behavior and Cognition," *European Journal of Human Genetics* 22, no. 6 (June 2014): 776–83, doi.org/10.1038/ejhg.2013.243.

49 In Italy, another boy: Maria Letizia Minniti et al., "Expanding the Phenotype of Brunner Syndrome from Childhood to Adulthood: Description of the Second Pediatric Patient and His Mother," *American Journal of Medical Genetics Part A* 194, no. 1 (2024): 82–87, doi.org/10.1002/ajmg.a.63413.

50 The scientists publishing the case report: E. E. Palmer et al., "New Insights into Brunner Syndrome and Potential for Targeted Therapy," *Clinical Genetics* 89, no. 1 (2016): 120–27, doi.org/10.1111/cge.12589.

51 Someone is affected by moral luck: Thomas Nagel, *Mortal Questions* (Cambridge and New York: Cambridge University Press, 1979).

52 *"You are poised":* Sophocles, *Antigone*, trans. Richard C. Jebb (Cambridge: Cambridge University Press, 1891), line 996, perseus.tufts.edu

/hopper/text?doc=Perseus%3Atext%3A1999.01.0186%3Acard%3D988. The line is spoken by the blind prophet Tiresias, whom Dante portrays in *The Inferno* as one of those condemned to walk around with his head on backward.

52 "She combines Z's atoms": Alfred R. Mele, *Manipulated Agents: A Window to Moral Responsibility* (Oxford and New York: Oxford University Press, 2019), 83. Mele's story further describes Ernie as "an ideally self-controlled agent who, in thirty years, will judge, on the basis of rational deliberation, that it is best to [perform an action, A] and will [perform an action, A] on the basis of that judgment, thereby bringing about E." The obvious disanalogy between the zygote/Ernie story and Brunner syndrome, then, is that men with Brunner syndrome are not easily characterized as self-controlled agents acting on the basis of rational deliberation, especially now that their behavior is known to have a genetic cause. Notably, however, several men who were later found to have Brunner syndrome were, prior to their diagnosis, found guilty of violent crimes and imprisoned. In the eyes of the legal system, then, they were treated as persons who were generally able to be guided by reason and who were thus responsible for their actions.

53 a Chinese scientist: Jon Cohen, "Did CRISPR Help—or Harm—the First-Ever Gene-Edited Babies?," *Science*, August 1, 2019, science.org/content/article/did-crispr-help-or-harm-first-ever-gene-edited-babies.

54 the free will and moral responsibility skeptics: Derk Pereboom, *Living Without Free Will* (Cambridge: Cambridge University Press, 2006).

55 "A manipulator may succeed": Harry Frankfurt, "Reply to John Martin Fischer," in *Contours of Agency: Essays on Themes from Harry Frankfurt*, ed. Sarah Buss and Lee Overton (Cambridge: MIT Press, 2002), 28, doi.org/10.7551/mitpress/2143.001.0001.

55 the residents of a village gather: Shirley Jackson, "'The Lottery,' by Shirley Jackson," *The New Yorker*, June 18, 1948, newyorker.com/magazine/1948/06/26/the-lottery.

56 the magazine received a flurry of mail: Ruth Franklin, "'The Lottery' Letters," *The New Yorker*, June 25, 2013, newyorker.com/books/page-turner/the-lottery-letters.

57 "developmental noise": Kevin J. Mitchell, "Developmental Noise Is an Overlooked Contributor to Innate Variation in Psychological Traits," *Behavioral and Brain Sciences* 45 (September 13, 2022): e171, doi.org/10.1017/S0140525X21001655.

58 "the power to constitute one's own being": Pagels, *Adam, Eve, and the Serpent*, 74.

58 imperial and ecclesiastical authority: Pagels, *Adam, Eve, and the Serpent*, 125, 145. She writes, "By insisting that humanity, ravaged by sin, now lies helplessly in need of outside intervention, Augustine's theory could not only validate secular power but justify as well the imposition of church authority—by force, if necessary—as essential for human salvation." She further quotes historian Peter Brown, who wrote that Pelagian ideas "cut at the root of episcopal authority. . . . By appeasing the

Pelagians the Catholic church would lose the vast authority it had begun to wield as the only force that could 'liberate' men from themselves." See Peter Brown, *Augustine of Hippo: A Biography* (Berkeley, 1969), 138.

Animal

61 **an intraperitoneal injection of thiopental:** Claude Messier, Sylvie Émond, and Katia Ethier, "New Techniques in Stereotaxic Surgery and Anesthesia in the Mouse," *Pharmacology Biochemistry and Behavior* 63, no. 2 (June 1, 1999): 313–18, doi.org/10.1016/S0091-3057(98)00247-0.

64 **The opioid system encodes:** K. C. Berridge, "Food Reward: Brain Substrates of Wanting and Liking," *Neuroscience and Biobehavioral Reviews* 20, no. 1 (1996): 1–25, doi.org/10.1016/0149-7634(95)00033-b; J. M. Delfs et al., "Noradrenaline in the Ventral Forebrain Is Critical for Opiate Withdrawal-Induced Aversion," *Nature* 403, no. 6768 (January 2000): 430–34, doi.org/10.1038/35000212.

65 **the country was awash in pills:** Patrick Radden Keefe, *Empire of Pain: The Secret History of the Sackler Dynasty* (New York: Doubleday, 2021).

65 **In 1997, the director:** Alan I. Leshner, "Addiction Is a Brain Disease, and It Matters," *Science* 278, no. 5335 (October 3, 1997): 45–47, doi.org/10.1126/science.278.5335.45.

66 **"What is natural":** St. Augustine, *Opus Imperfectum Contra Julianum*, as cited in Pagels, *Adam, Eve, and the Serpent*, 135.

66 **"If others didn't share":** Judith Grisel, *Never Enough: The Neuroscience and Experience of Addiction* (New York: Doubleday, 2019), 6–7.

67 **"Today, we are learning":** "President Clinton: Announcing the Completion of the First Survey of the Entire Human Genome, June 26, 2000," accessed May 3, 2024, clintonwhitehouse3.archives.gov/WH/Work/062600.html.

67 **three times more likely to abuse drugs:** Kenneth S. Kendler et al., "An Extended Swedish National Adoption Study of Alcohol Use Disorder," *JAMA Psychiatry* 72, no. 3 (March 2015): 211–18, doi.org/10.1001/jamapsychiatry.2014.2138; Kenneth S. Kendler et al., "Genetic and Familial Environmental Influences on the Risk for Drug Abuse: A National Swedish Adoption Study," *Archives of General Psychiatry* 69, no. 7 (July 1, 2012): 690–97, doi.org/10.1001/archgenpsychiatry.2011.2112.

67 **Identical twins are as similar:** Kenneth S. Kendler et al., "Genetic and Family and Community Environmental Effects on Drug Abuse in Adolescence: A Swedish National Twin and Sibling Study," *American Journal of Psychiatry* 171, no. 2 (February 2014): 209–17, doi.org/10.1176/appi.ajp.2013.12101300.

67 **Drug abuse was shown to run in families:** Kenneth S. Kendler et al., "The Patterns of Family Genetic Risk Scores for Eleven Major Psychiatric and Substance Use Disorders in a Swedish National Sample," *Translational Psychiatry* 11, no. 1 (May 27, 2021): 1–8, doi.org/10.1038/s41398-021-01454-z.

68 **about 50 percent of the differences:** A. Agrawal et al., "The Genetics of

Addicton—a Translational Perspective," *Translational Psychiatry* 2, no. 7 (July 2012): e140, doi.org/10.1038/tp.2012.54.

68 **"evidence from adoption"**: Alan I. Leshner, "Vulnerability to Addiction: New Research Opportunities," *American Journal of Medical Genetics* 96, no.5(2000):590–91,doi.org/10.1002/1096-8628(20001009)96:5<590::AID-AJMG2>3.0.CO;2-2.

69 **"numbering, chief of all"**: Aeschylus, *Prometheus Bound*, trans. Joel Agee (New York: NYRB Classics, 2015); Martha C. Nussbaum, *The Fragility of Goodness: Luck and Ethics in Greek Tragedy and Philosophy* (Cambridge and New York: Cambridge University Press, 2001).

70 **Fighting fish (*Betta splendens*)**: Wanchang Zhang et al., "The Genetic Architecture of Phenotypic Diversity in the Betta Fish (Betta Splendens)," *Science Advances* 8, no. 38 (September 21, 2022): eabm4955, doi.org/10.1126/sciadv.abm4955.

70 **domesticating Siberian silver foxes**: Lee Alan Dugatkin, "The Silver Fox Domestication Experiment," *Evolution: Education and Outreach* 11, no. 1 (December 7, 2018): 16, doi.org/10.1186/s12052-018-0090-x.

70 **The BALB/cJ mouse**: Maze Engineers, "BALB/cJ Mouse Strain," *Maze Engineers* (blog), November 24, 2018, conductscience.com/maze/balb-cj-mouse-strain.

70 **genetically altered one-cell mouse embryos**: Olivier Cases et al., "Aggressive Behavior and Altered Amounts of Brain Serotonin and Norepinephrine in Mice Lacking MAOA," *Science* 268, no. 5218 (June 23, 1995): 1763–66, doi.org/10.1126/science.7792602.

71 **bit more often**: Anna L. Scott et al., "Novel Monoamine Oxidase A Knock Out Mice with Human-like Spontaneous Mutation," *Neuroreport* 19, no. 7 (May 7, 2008): 739–43, doi.org/10.1097/WNR.0b013e3282fd6e88.

71 **"axioms in philosophy"**: "Letter #69: To John Hamilton Reynolds, 3 May 1818," *The Keats Letters Project* (blog), May 3, 2018, keatslettersproject.com/letters/letter-69-to-john-hamilton-reynolds-3-may-1818.

72 **reduce their cocaine use**: Amber N. Edinoff et al., "Cebranopadol for the Treatment of Chronic Pain," *Current Pain and Headache Reports* 27, no. 10 (October 1, 2023): 615–22, doi.org/10.1007/s11916-023-01148-9; Giordano de Guglielmo et al., "Cebranopadol Blocks the Escalation of Cocaine Intake and Conditioned Reinstatement of Cocaine Seeking in Rats," *Journal of Pharmacology and Experimental Therapeutics* 362, no. 3 (September 1, 2017): 378–84, doi.org/10.1124/jpet.117.241042.

73 **"circumvents denial of the flesh"**: Cathy Majtenyi, "Quick Fix Weight Loss Denies the Cross," *Catholic Register*, April 13, 2023, catholicregister.org/opinion/guest-columnists/item/35437-quick-fix-weight-loss-denies-the-cross.

73 **40 percent less likely to overdose**: Fares Qeadan, Ashlie McCunn, and Benjamin Tingey, "The Association Between Glucose-Dependent Insulinotropic Polypeptide and/or Glucagon-like Peptide-1 Receptor Agonist Prescriptions and Substance-Related Outcomes in Patients with Opioid and Alcohol Use Disorders: A Real-World Data Analysis," *Addiction* 120, no. 2 (2025): 236–50, doi.org/10.1111/add.16679.

73 **pharmaceutical executives considered weight**: Gina Kolata, "We Know

Where New Weight Loss Drugs Came From, but Not Why They Work," *The New York Times*, August 17, 2023, nytimes.com/2023/08/17/health/weight-loss-drugs-obesity-ozempic-wegovy.html.

74 **In *Paradise Lost*:** John Milton, *Paradise Lost*, ed. William Kerrigan, John Rumrich, and Stephen M. Fallon (New York: Modern Library, 2008).

74 **another reading of Genesis:** Howard Nemerov, "Cain (a New Play)," *Tulane Drama Review* 4, no. 2 (December 1959): 12–26, doi.org/10.2307/1124858. Nemerov had a strong rivalry with his own sibling, photographer Diane Arbus, whose most well-known work is a photograph of identical twin girls. See Larissa Archer, "The Sibling Rivalry That Shaped Diane Arbus's Vision," *Hyperallergic* (March 27, 2015), hyperallergic.com/194141/the-sibling-rivalry-that-shaped-diane-arbuss-vision/; Alexander Nemerov, *Silent Dialogues: Diane Arbus and Howard Nemerov* (San Francisco: Fraenkel Gallery, 2015).

75 **Landrigan's problems emerged:** "Murder Gene," *60 Minutes II*, Columbia Broadcasting System, 2001, video.alexanderstreet.com/watch/murder-gene-2.

75 **As a young man:** "Schriro v. Landrigan," *Harvard Law Review* 121, no. 1 (November 2007): 255–65.

75 **Darrell's own father:** "Murder Gene."

76 **"out of the ordinary":** *Arizona v. Landrigan*, 176 Ariz. 1 (1993); *Landrigan v. Stewart*, 272 F.3d 1221 (9th Cir. 2001).

76 **conduct disorder:** American Psychiatric Association, *Diagnostic and Statistical Manual of Mental Disorders, Text Revision DSM-5-TR*.

76 **Boys are diagnosed at about twice the rate of girls:** Graeme Fairchild et al., "Conduct Disorder," *Nature Reviews Disease Primers* 5, no. 1 (June 27, 2019): 1–25, doi.org/10.1038/s41572-019-0095-y.

76 **generally similar across racial and ethnic groups:** Joan McCord, Cathy Spatz Widom, and Nancy A. Crowell, eds., *Juvenile Crime, Juvenile Justice* (Washington, D.C.: National Academy Press, 2001), doi.org/10.17226/9747.

77 **"callous-unemotionality":** Meagan Docherty et al., "Developmental Trajectories of Interpersonal Callousness from Childhood to Adolescence as Predictors of Antisocial Behavior and Psychopathic Features in Young Adulthood," *Journal of Abnormal Psychology* 128, no. 7 (2019): 700–709, doi.org/10.1037/abn0000449; Terrie E. Moffitt, "A Review of Research on the Taxonomy of Life-Course Persistent versus Adolescence-Limited Antisocial Behavior," in *The Cambridge Handbook of Violent Behavior and Aggression* (New York: Cambridge University Press, 2007), 49–74, doi.org/10.1017/CBO9780511816840.004; T. E. Moffitt, "Adolescence-Limited and Life-Course-Persistent Antisocial Behavior: A Developmental Taxonomy," *Psychological Review* 100, no. 4 (October 1993): 674–701.

77 **"dark triad":** Chris Patrick, "Psychopathy," *Noba Textbook Series*, 2025, nobaproject.com/modules/psychopathy.

78 **One study of forty-nine countries:** Peter K. Jonason et al., "Country-Level Correlates of the Dark Triad Traits in 49 Countries," *Journal of Personality* 88, no. 6 (2020): 1252–67, doi.org/10.1111/jopy.12569.

78 **"Not all psychopaths":** Robert D. Hare, "The Predators Among Us," key-

note address, Canadian Police Association Annual Meeting, St. John's, Newfoundland and Labrador, August 2002.
78 **engagement in unethical behaviors:** Ernest H. O'Boyle Jr. et al., "A Meta-Analysis of the Dark Triad and Work Behavior: A Social Exchange Perspective," *Journal of Applied Psychology* 97, no. 3 (2012): 557–79, doi.org/10.1037/a0025679; M. Gholami et al., "From Dark Triad Personality Traits to Digital Harm: Mediating Cyberbullying Through Online Moral Disengagement," *Deviant Behavior*, 2025, doi.org/10.1080/01639625.2025.2453445.
79 **Adopted children whose biological fathers:** T. Frisell, P. Lichtenstein, and N. Långström, "Violent Crime Runs in Families: A Total Population Study of 12.5 Million Individuals," *Psychological Medicine* 41, no. 1 (January 2011): 97–105, doi.org/10.1017/S0033291710000462; K. S. Kendler et al., "A Swedish National Adoption Study of Criminality," *Psychological Medicine* 44, no. 9 (July 2014): 1913–25, doi.org/10.1017/S0033291713002638.
80 **about 80 percent of the differences:** Irving I. Gottesman and James Shields, "A Critical Review of Recent Adoption, Twin, and Family Studies of Schizophrenia: Behavioral Genetics Perspectives," *Schizophrenia Bulletin* 2, no. 3 (1976): 360–401, doi.org/10.1093/schbul/2.3.360.
80 **the most severe form of child conduct:** Essi Viding et al., "Evidence for Substantial Genetic Risk for Psychopathy in 7-Year-Olds," *Journal of Child Psychology and Psychiatry and Allied Disciplines* 46, no. 6 (June 2005): 592–97, doi.org/10.1111/j.1469-7610.2004.00393.x.
80 **antisocial behaviors in adulthood:** Arianna M. Gard, Hailey L. Dotterer, and Luke W. Hyde, "Genetic Influences on Antisocial Behavior: Recent Advances and Future Directions," *Current Opinion in Psychology* 27 (June 1, 2019): 46–55, doi.org/10.1016/j.copsyc.2018.07.013; S. Alexandra Burt, "The Genetic, Environmental, and Cultural Forces Influencing Youth Antisocial Behavior Are Tightly Intertwined," *Annual Review of Clinical Psychology* 18, no. 1 (2022): 155–78, doi.org/10.1146/annurev-clinpsy-072220-015507.
80 **not specific to opioids:** Mark R. Jones et al., "A Brief History of the Opioid Epidemic and Strategies for Pain Medicine," *Pain and Therapy* 7, no. 1 (June 2018): 13–21, doi.org/10.1007/s40122-018-0097-6.
80 ***CADM2* is associated with:** Joëlle A. Pasman et al., "The CADM2 Gene and Behavior: A Phenome-Wide Scan in UK-Biobank," *Behavior Genetics* 52, no. 4 (September 1, 2022): 306–14, doi.org/10.1007/s10519-022-10109-8.
81 **a disorder of speech and language:** Cecilia S. L. Lai et al., "A Forkhead-Domain Gene Is Mutated in a Severe Speech and Language Disorder," *Nature* 413, no. 6855 (October 2001): 519–23, doi.org/10.1038/35097076.
81 **opioid use disorder:** Joseph D. Deak et al., "Genome-Wide Association Study in Individuals of European and African Ancestry and Multi-Trait Analysis of Opioid Use Disorder Identifies 19 Independent Genome-Wide Significant Risk Loci," *Molecular Psychiatry* 27, no. 10 (October 2022): 3970–79, doi.org/10.1038/s41380-022-01709-1.
81 **cannabis use:** Emma C. Johnson et al., "A Large-Scale Genome-Wide

Association Study Meta-Analysis of Cannabis Use Disorder," *The Lancet Psychiatry* 7, no. 12 (December 1, 2020): 1032–45, doi.org/10.1016/S2215-0366(20)30339-4.

81 **symptoms of ADHD:** Dorothy Clyde, "Risk Loci for ADHD," *Nature Reviews Genetics* 20, no. 2 (February 2019): 69, doi.org/10.1038/s41576-018-0084-0.

81 **As do aggressive people:** Jorim J. Tielbeek et al., "Uncovering the Genetic Architecture of Broad Antisocial Behavior Through a Genome-Wide Association Study Meta-Analysis," *Molecular Psychiatry* 27, no. 11 (November 2022): 4453–63, doi.org/10.1038/s41380-022-01793-3.

81 **we keep identifying more and more genes:** Karlsson Linnér et al., "Multivariate Analysis of 1.5 Million People Identifies Genetic Associations with Traits Related to Self-Regulation and Addiction"; Alexander S. Hatoum et al., "Multivariate Genome-Wide Association Meta-Analysis of over 1 Million Subjects Identifies Loci Underlying Multiple Substance Use Disorders," *Nature Mental Health* 1, no. 3 (March 2023): 210–23, doi.org/10.1038/s44220-023-00034-y.

82 **"Grain upon grain":** Samuel Beckett, *Endgame: A Play by Samuel Beckett* (New York: Grove, 1981).

82 **adding up all the information:** Daniel W. Belsky and K. Paige Harden, "Phenotypic Annotation: Using Polygenic Scores to Translate Discoveries from Genome-Wide Association Studies from the Top Down," *Current Directions in Psychological Science* 28, no. 1 (2019): 82–90.

82 **"polygenic" correlations are about as large:** Peter T. Tanksley et al., "Do Polygenic Indices Capture 'Direct' Effects on Child Externalizing Behavior Problems? Within-Family Analyses in Two Longitudinal Birth Cohorts," *Clinical Psychological Science*, August 24, 2024, doi.org/10.1177/21677026241260260.

84 **"I personally think":** "Murder Gene."

84 **"It is highly doubtful":** *Landrigan v. Stewart*, 272 F.3d. (9th Cir. 2001).

84 **the Supreme Court agreed:** *Schriro v. Landrigan*, 550 U.S. 465 (2007); "Schriro v. Landrigan," *Oyez*, oyez.org/cases/2006/05-1575.

84 **Inheritance, the court thought:** Azgad Gold and Paul S. Appelbaum, "The Inclination to Evil and the Punishment of Crime—from the Bible to Behavioral Genetics," *Israel Journal of Psychiatry and Related Sciences* 51, no. 3 (2014): 162–68.

85 **Jeffrey Landrigan made legal headlines:** Chris McGreal, "Arizona Execution Goes Ahead After Stay Lifted," *The Guardian*, October 27, 2010, theguardian.com/world/2010/oct/27/arizona-execution-stay-lifted.

85 **Landrigan died by lethal injection:** John Schwartz, "Murderer Executed in Arizona," *The New York Times*, October 27, 2010, nytimes.com/2010/10/28/us/28execute.html.

Choice

86 **The Rotunda was constructed:** Brendan Wolfe, "Slavery at the University of Virginia," *Encyclopedia Virginia*, Virginia Humanities, Decem-

ber 7, 2020, encyclopediavirginia.org/entries/slavery-at-the-university-of-virginia.
86 "the illimitable freedom": Andrew J. O'Shaughnessy and Edward L. Ayers, *The Illimitable Freedom of the Human Mind: Thomas Jefferson's Idea of a University* (Charlottesville: University of Virginia Press, 2021).
88 "The essential source": Thomas Nagel, *The View from Nowhere* (New York and London: Oxford University Press, 1989).
89 **Ted Chiang's sci-fi novella**: Ted Chiang, *Exhalation: Stories* (New York: Knopf, 2019). The title of Chiang's novella, "Anxiety Is the Dizziness of Freedom," is taken from Søren Kierkegaard, *The Concept of Anxiety: A Simple Psychologically Orienting Deliberation on the Dogmatic Issue of Hereditary Sin*, Kierkegaard's Writings, vol. 8 (Princeton: Princeton University Press, 2013 [1844]), doi.org/10.1515/9781400846979.
89 A main character in the story: Adam Phillips, *Missing Out: In Praise of the Unlived Life* (New York: Picador, 2013).
91 **curate data from every twin study**: Tinca J. C. Polderman et al., "Meta-Analysis of the Heritability of Human Traits Based on Fifty Years of Twin Studies," *Nature Genetics* 47, no. 7 (July 2015): 702–9, doi.org/10.1038/ng.3285.
91 **Nazis were obsessed with twins**: Erin Blakemore, "Why the Nazis Were Obsessed with Twins," *History*, July 8, 2019, history.com/news/nazi-twin-experiments-mengele-eugenics.
91 **Far-right reactionaries still twist**: Daniel J. Kevles, *In the Name of Eugenics: Genetics and the Uses of Human Heredity* (Cambridge: Harvard University Press, 1998); Adam Rutherford, *How to Argue with a Racist* (London: Weidenfeld & Nicolson, 2020); Harden, *Genetic Lottery*.
92 **Twin studies make**: Robert Plomin et al., *Behavioral Genetics* (New York: Worth Publishers, 2012); Marcus W. Feldman and Sohini Ramachandran, "Missing Compared to What? Revisiting Heritability, Genes, and Culture," *Philosophical Transactions of the Royal Society of London, Series B, Biological Sciences* 373, no. 1743 (April 5, 2018): 20170064, doi.org/10.1098/rstb.2017.0064; Dalton Conley et al., "Heritability and the Equal Environments Assumption: Evidence from Multiple Samples of Misclassified Twins," *Behavior Genetics* 43, no. 5 (September 2013): 415–26, doi.org/10.1007/s10519-013-9602-1.
92 **bottomless question of nature versus nurture**: Eric Turkheimer, "Three Laws of Behavior Genetics and What They Mean," *Current Directions in Psychological Science*, October 1, 2000, journals.sagepub.com/doi/10.1111/1467-8721.00084.
92 **twins whose lives had turned out unalike**: Eric Turkheimer and K. Paige Harden, "Behavior Genetic Research Methods: Testing Quasi-Causal Hypotheses Using Multivariate Twin Data," in *Handbook of Research Methods in Social and Personality Psychology*, 2nd ed. (New York: Cambridge University Press, 2014), 159–87.
92 **dissertation on pairs of identical twin girls**: Mary Christina Waldron, "Parsing Quasi-Causal Relations from Confounds: A Study of Teenage Childbearing in Australian Twins," PhD dissertation, University of Vir-

ginia, 2004, proquest.com/docview/305106742/abstract/1503271549834D4FPQ/1.

92 **Waldron discovered that:** Bodine M. A. Gonggrijp et al., "The Co-Twin Control Design: Implementation and Methodological Considerations," *Twin Research and Human Genetics* 26, no. 4–5 (August 2023): 249–56, doi.org/10.1017/thg.2023.35.

92 **Minnesota Study of Twins Reared Apart:** T. J. Bouchard et al., "Sources of Human Psychological Differences: The Minnesota Study of Twins Reared Apart," *Science* 250, no. 4978 (October 12, 1990): 223–28, doi.org/10.1126/science.2218526; Nancy L. Segal, *Born Together—Reared Apart: The Landmark Minnesota Twin Study* (Cambridge: Harvard University Press, 2012), doi.org/10.4159/harvard.9780674065154.

93 **partially funded by the Pioneer Fund:** Paul Lombardo, "'The American Breed': Nazi Eugenics and the Origin of the Pioneer Fund," *Albany Law Review* 65, no. 3 (January 1, 2002): 743–830. In addition to receiving money from the Pioneer Fund, MISTRA also regularly received funding from Koch Industries, one of the largest privately held companies in the United States, which was founded by Fred C. Koch in 1940. Koch hired a German nanny, an ardent Nazi, to look after his children, and wrote that "the only sound countries in the world are Germany, Italy, and Japan." Koch Industries became profitable by building an oil refinery for the Nazi regime. Jane Mayer, *Dark Money: The Hidden History of the Billionaires Behind the Rise of the Radical Right* (New York: Doubleday, 2016).

94 **they grew ever more like each other:** Elliot M. Tucker-Drob, Daniel A. Briley, and K. Paige Harden, "Genetic and Environmental Influences on Cognition Across Development and Context," *Current Directions in Psychological Science* 22, no. 5 (October 1, 2013): 349–55, doi.org/10.1177/0963721413485087; Daniel A. Briley and Elliot M. Tucker-Drob, "Comparing the Developmental Genetics of Cognition and Personality over the Life Span," *Journal of Personality* 85, no. 1 (2017): 51–64, doi.org/10.1111/jopy.12186; Sarah E. Bergen, Charles O. Gardner, and Kenneth S. Kendler, "Age-Related Changes in Heritability of Behavioral Phenotypes over Adolescence and Young Adulthood: A Meta-Analysis," *Twin Research and Human Genetics: The Official Journal of the International Society for Twin Studies* 10, no. 3 (June 2007): 423–33, doi.org/10.1375/twin.10.3.423; Bouchard et al., "Sources of Human Psychological Differences"; Segal, *Born Together—Reared Apart*.

94 **"If everything we call":** Arthur Allen, "Nature and Nurture: When It Comes to Twins, Sometimes It's Hard to Tell the Two Apart," *The Washington Post*, January 11, 1998, washingtonpost.com/wp-srv/national/longterm/twins/twins2.htm.

94 **how tall they are:** Karri Silventoinen et al., "Heritability of Adult Body Height: A Comparative Study of Twin Cohorts in Eight Countries," *Twin Research and Human Genetics* 6, no. 5 (October 2003): 399–408, doi.org/10.1375/twin.6.5.399.

94 **their body mass index:** R. Arlen Price and Irving I. Gottesman, "Body Fat in Identical Twins Reared Apart: Roles for Genes and Environment,"

Behavior Genetics 21, no. 1 (January 1, 1991): 1–7, doi.org/10.1007/BF01067662.

94 **religious fundamentalism and religious values:** Niels G. Waller et al., "Genetic and Environmental Influences on Religious Interests, Attitudes, and Values: A Study of Twins Reared Apart and Together," *Psychological Science* 1 (1990): 138–42, doi.org/10.1111/j.1467-9280.1990.tb00083.x.

94 **ended up being similar:** Laura B. Koenig and Matt McGue, "The Behavioral Genetics of Religiousness," *Theology and Science* 9, no. 2 (May 1, 2011): 199–212, doi.org/10.1080/14746700.2011.563585.

95 **genetic influence on religiousness goes up:** Koenig and McGue, "Behavioral Genetics of Religiousness."

96 **"Science deals with humans":** Allen, "Nature and Nurture."

96 **"A Behavioral Genetic Investigation":** Kathryn Paige Harden, "A Behavioral Genetic Investigation of Religiosity and Adolescent Problem Behavior," PhD dissertation, University of Virginia, 2009.

98 **In the early 1980s, neuroscientists:** B. Libet et al., "Time of Conscious Intention to Act in Relation to Onset of Cerebral Activity (Readiness-Potential): The Unconscious Initiation of a Freely Voluntary Act," *Brain: A Journal of Neurology* 106 (Pt. 3) (September 1983): 623–42, doi.org/10.1093/brain/106.3.623.

99 **"Be sure of this":** Friedrich Nietzsche, *The Dawn of Day*, trans. John McFarland Kennedy (New York: Macmillan, 1911), gutenberg.org/files/39955/39955-h/39955-h.html.

99 **sensorimotor stage of development:** Robert Siegler, "Cognitive Development in Childhood," *Noba Textbook Series*, 2019, accessed March 17, 2025, nobaproject.com/modules/cognitive-development-in-childhood.

100 **A later study used:** Uri Maoz et al., "Neural Precursors of Decisions That Matter—an ERP Study of Deliberate and Arbitrary Choice," ed. Redmond G. O'Connell et al., *eLife* 8 (October 23, 2019): e39787, doi.org/10.7554/eLife.39787.

101 **In "Cain," the Nemerov poem:** Nemerov, "Cain (a New Play)."

103 **Science finds evidence of genetic influences:** Turkheimer, "Three Laws of Behavior Genetics and What They Mean"; Polderman et al., "Meta-Analysis of the Heritability of Human Traits Based on Fifty Years of Twin Studies."

103 **gored its owners twice:** "Act Two: If By Chance We Meet Again," *This American Life*, December 12, 2017, thisamericanlife.org/291/reunited-and-it-feels-so-good/act-two-6.

105 **sinister reputation as the enemy of freedom:** The phrase "sinister reputation" is taken from Richard Dawkins, *The Extended Phenotype: The Long Reach of the Gene* (Oxford: Oxford University Press, 2016).

105 **One of the most influential books:** St. Augustine, *The Confessions*, trans. Maria Boulding (Hyde Park, New York: New City Press, 1997).

105 **"Anyone who had examined":** St. Augustine, *Confessions*, VII.10.

105 **"torn and wounded":** St. Augustine, *Confessions*, VI.25.

106 **"A huge storm blew up":** St. Augustine, *Confessions*, VIII.28.

106 **"Pick it up and read":** St. Augustine, *Confessions*, VIII.29.

106 "I was shackled": St. Augustine, *Confessions*, VI.21.
106 "By the just retribution": St. Augustine, *The City of God*, vol. 2, trans. Marcus Dods (New York: Hafner, 1948), XIV.15, 30.
107 "The flesh lusts": St. Augustine, *Confessions*, VIII.11.
109 ordained as an Episcopal priest: Matthew J. Milliner, "Hive Mind: Alan Watts, Thomas Merton, and the Church of the East," *The Hedgehog Review*, Fall 2024, hedgehogreview.com/issues/in-need-of-repair/articles/hive-mind.
109 "The serpent didn't make": *Alan Watts—The Present Moment (Boat Analogy)*, 2014, youtube.com/watch?v=fAsHGEB69P0.
110 we give our actions Necessity's name: Sophie-Grace Chappell, "Agamemnon at Aulis: On the Right and Wrong Sorts of Imaginative Identification," *Topoi*, May 18, 2023, doi.org/10.1007/s11245-023-09920-7.
111 portrayed as morally objectionable: Nussbaum, *Fragility of Goodness*.
111 "We smelt off him": Chappell, "Agamemnon at Aulis."

Essence

112 Robert Harris was born: Gary Watson, "Responsibility and the Limits of Evil: Variations on a Strawsonian Theme," in *Agency and Answerability: Selected Essays*, ed. Gary Watson (Oxford University Press, 2004), doi.org/10.1093/acprof:oso/9780199272273.003.0009.
112 elevates one's own risk for violence: Kendler et al., "Swedish National Adoption Study of Criminality."
112 Children who were born early: Samantha Johnson and Neil Marlow, "Preterm Birth and Childhood Psychiatric Disorders," *Pediatric Research* 69, no. 8 (May 2011): 11–18, doi.org/10.1203/PDR.0b013e318212faa0; Geethanjali Lingasubramanian et al., "Gestational Age at Term and Teacher-Reported Attention-Deficit Hyperactivity Disorder Symptom Patterns," *Journal of Pediatrics* 251 (December 2022): 120–26, doi.org/10.1016/j.jpeds.2022.07.042.
112 heavily exposed to alcohol in the womb: Mayo Clinic, "Fetal Alcohol Syndrome—Symptoms and Causes," accessed February 20, 2025, mayoclinic.org/diseases-conditions/fetal-alcohol-syndrome/symptoms-causes/syc-20352901.
112 Even rats who are exposed to alcohol: Sandra J. Kelly, Nancy Day, and Ann P. Streissguth, "Effects of Prenatal Alcohol Exposure on Social Behavior in Humans and Other Species," *Neurotoxicology and Teratology* 22, no. 2 (2000): 143–49. Notably, alterations in social behavior among alcohol-exposed rodents persist to adulthood and include maternal behaviors. The authors described the behavior of rodent mothers who had been exposed to alcohol prenatally as "disoriented, unmotivated, and disorganized. . . . They spent much time grooming themselves and eating and drinking while ignoring the pups."
113 Particularly affected is the cerebellum: Eric C. H. Leung et al., "Recent Breakthroughs in Understanding the Cerebellum's Role in Fetal Alcohol Spectrum Disorder: A Systematic Review," *Alcohol*, December 13, 2023, doi.org/10.1016/j.alcohol.2023.12.003.

113 "the mind's quality control center": Scott Marek et al., "Spatial and Temporal Organization of the Individual Human Cerebellum," *Neuron* 100, no. 4 (2018): 977–93, doi.org/10.1016/j.neuron.2018.10.010.

113 important in genetic studies: Karlsson Linnér et al., "Multivariate Analysis of 1.5 Million People Identifies Genetic Associations with Traits Related to Self-Regulation and Addiction."

113 the pathways of risk: Burt, "The Genetic, Environmental, and Cultural Forces Influencing Youth Antisocial Behavior Are Tightly Intertwined."

113 "He'd come up to": Watson, "Responsibility and the Limits of Evil."

113 a remarkable study: Avshalom Caspi et al., "Maternal Expressed Emotion Predicts Children's Antisocial Behavior Problems: Using Monozygotic-Twin Differences to Identify Environmental Effects on Behavioral Development," *Developmental Psychology* 40, no. 2 (2004): 149–61, doi.org/10.1037/0012-1649.40.2.149.

114 evidence of "child effects": David Reiss et al., "Parenting in the Context of the Child: Genetic and Social Processes," *Monographs of the Society for Research in Child Development* 87, no. 1–3 (2022): 7–188, doi.org/10.1111/mono.12460.

114 This callousness, in turn, predicts: Luke W. Hyde et al., "Heritable and Non-Heritable Pathways to Early Callous-Unemotional Behaviors," *American Journal of Psychiatry* 173, no. 9 (September 1, 2016): 903–10, doi.org/10.1176/appi.ajp.2016.15111381.

115 the "good enough" mother: D. W. Winnicott, *Playing and Reality* (London: Routledge, 2005).

115 demonstrated just how crucial: Deborah Blum, *Love at Goon Park: Harry Harlow and the Science of Affection* (Cambridge: Perseus Publishing, 2002); Harry F. Harlow, "The Nature of Love," *American Psychologist* 13, no. 12 (1958): 673–85, doi.org/10.1037/h0047884; Lenny van Rosmalen, Maartje P.C.M. Luijk, and Frank C. P. van der Horst, "Harry Harlow's Pit of Despair: Depression in Monkeys and Men," *Journal of the History of the Behavioral Sciences* 58, no. 2 (2022): 204–22, doi.org/10.1002/jhbs.22180.

116 more anxious and more aggressive: Carine I. Parent and Michael J. Meaney, "The Influence of Natural Variations in Maternal Care on Play Fighting in the Rat," *Developmental Psychobiology* 50, no. 8 (2008): 767–76, doi.org/10.1002/dev.20342.

116 *NR3C1* gene hyper-methylation in children: Dante Cicchetti and Elizabeth D. Handley, "Methylation of the Glucocorticoid Receptor Gene (NR3C1) in Maltreated and Nonmaltreated Children: Associations with Behavioral Undercontrol, Emotional Lability/Negativity, and Externalizing and Internalizing Symptoms," *Development and Psychopathology* 29, no. 5 (December 2017): 1795–1806, doi.org/10.1017/S0954579417001407.

116 *NR3C1* DNA methylation in their hippocampus: Patrick O. McGowan et al., "Epigenetic Regulation of the Glucocorticoid Receptor in Human Brain Associates with Childhood Abuse," *Nature Neuroscience* 12, no. 3 (March 2009): 342–48, doi.org/10.1038/nn.2270.

116 "It is better to be a sinner": W.R.D. Fairbairn, *Psychoanalytic Studies of the Personality* (London and New York: Routledge, 1994).

118 **shocked to recognize himself:** Dan Morain, "Son of Executed Killer Faces Trial: It Was a Shock for Robert Alton Harris Jr. to Learn the Identity of His Biological Father. Now, He Could Get 30 Years in Prison If Convicted of Robbery," *Los Angeles Times*, September 6, 1992, latimes.com/archives/la-xpm-1992-09-06-mn-287-story.html.

118 **robbed a cabdriver at gunpoint:** "Robert Harris Jr. Pleads Guilty in Robbery," *Los Angeles Times*, October 9, 1992, latimes.com/archives/la-xpm-1992-10-09-me-603-story.html.

118 **"a devil, a born devil":** William Shakespeare, *The Tempest*, ed. Barbara A. Mowat and Paul Werstine (New York: Simon & Schuster, 2004).

119 **The space between nature and nurture:** Nagel, *Mortal Questions*.

119 **"Genetical science has outgrown":** Evelyn Fox Keller, "Goodbye, Nature vs Nurture," *New Scientist*, September 15, 2010, newscientist.com/article/mg20727780-800-goodbye-nature-vs-nurture.

119 **"so-called heredity-environment question":** Anne Anastasi, "Heredity, Environment, and the Question 'How?,'" *Psychological Review* 65, no. 4 (1958): 197–208, doi.org/10.1037/h0044895.

119 **"was more due to his genes":** David T. Lykken, "The Genetics of Genius," in *Genius and Mind: Studies of Creativity and Temperament* (New York: Oxford University Press, 1998), 15–37, doi.org/10.1093/acprof:oso/9780198523734.003.0002.

119 **"We now know":** Evelyn Fox Keller, *The Mirage of a Space Between Nature and Nurture* (Durham, N.C.: Duke University Press, 2010).

120 **whispers in the scientific community:** Hewitt, "Editorial Policy on Candidate Gene Association and Candidate Gene-by-Environment Interaction Studies of Complex Traits."

121 **Children genetically predisposed:** Elliot M. Tucker-Drob and K. Paige Harden, "Intellectual Interest Mediates Gene × Socioeconomic Status Interaction on Adolescent Academic Achievement: Intellectual Interest and G×E," *Child Development*, January 2012, doi.org/10.1111/j.1467-8624.2011.01721.x; Tucker-Drob, Briley, and Harden, "Genetic and Environmental Influences on Cognition across Development and Context"; Elliot M. Tucker-Drob et al., "Emergence of a Gene × Socioeconomic Status Interaction on Infant Mental Ability Between 10 Months and 2 Years," *Psychological Science* 22, no. 1 (January 2011): 125–33, doi.org/10.1177/0956797610392926.

121 **Adolescents who inherit genes:** Natalie Kretsch, Jane Mendle, and K. Paige Harden, "A Twin Study of Objective and Subjective Pubertal Timing and Peer Influence on Risk-Taking," *Journal of Research on Adolescence: The Official Journal of the Society for Research on Adolescence* 26, no. 1 (March 1, 2016): 45–59, doi.org/10.1111/jora.12160.

121 **Adolescents with elevated testosterone levels:** Andrew D. Grotzinger et al., "Hair and Salivary Testosterone, Hair Cortisol, and Externalizing Behaviors in Adolescents," *Psychological Science* 29, no. 5 (May 1, 2018): 688–99, doi.org/10.1177/0956797617742981; Frank D. Mann et al., "Person × Environment Interactions on Adolescent Delinquency: Sensation Seeking, Peer Deviance, and Parental Monitoring," *Personality and Individual Differences* 76 (April 1, 2015): 129–34, doi.org/10.1016/j.paid.2014.11.055.

122 "What would be so specially bad": Daniel C. Dennett, *Freedom Evolves* (London: Penguin Publishing Group, 2004).

122 "Genetic causes and environmental causes": Dawkins, *Extended Phenotype*.

122 In some studies, participants appear: Kathryn Tabb, Matthew S. Lebowitz, and Paul S. Appelbaum, "Behavioral Genetics and Attributions of Moral Responsibility," *Behavior Genetics* 49, no. 2 (March 1, 2019): 128–35, doi.org/10.1007/s10519-018-9916-0; Matthew S. Lebowitz, Kathryn Tabb, and Paul S. Appelbaum, "Asymmetrical Genetic Attributions for Prosocial versus Antisocial Behaviour," *Nature Human Behaviour* 3, no. 9 (September 2019): 940–49, doi.org/10.1038/s41562-019-0651-1.

123 In one study, the psychiatrist Paul Appelbaum: Paul S. Appelbaum, Nicholas Scurich, and Raymond Raad, "Effects of Behavioral Genetic Evidence on Perceptions of Criminal Responsibility and Appropriate Punishment," *Psychology, Public Policy, and Law* 21, no. 2 (May 2015): 134–44, doi.org/10.1037/law0000039.supp.

123 Subsequent experiments replicated: Appelbaum, Scurich, and Raad, "Effects of Behavioral Genetic Evidence on Perceptions of Criminal Responsibility and Appropriate Punishment."

124 experiments led by the psychologist Susan Gelman: Meredith Meyer et al., "Genetic Essentialist Beliefs About Criminality Predict Harshness of Recommended Punishment," *Journal of Experimental Psychology: General* 151, no. 12 (December 2022): 3230–48, doi.org/10.1037/xge0001240.

124 most attempts to introduce genetic evidence: Deborah W. Denno, "Courts' Increasing Consideration of Behavioral Genetics Evidence in Criminal Cases: Results of a Longitudinal Study," SSRN Scholarly Paper, Social Science Research Network, Rochester, N.Y., 2011, papers.ssrn.com/abstract=2065523; Deborah W. Denno, "Behavioral Genetics Evidence in Criminal Cases: 1994–2007," SSRN Scholarly Paper, Social Science Research Network, Rochester, N.Y., March 1, 2009, papers.ssrn.com/abstract=1089171; Denno, "What Real-World Criminal Cases Tell Us About Genetics Evidence."

124 "mitigating factors": *Woodson v. North Carolina,* 428 U.S. 280 (1976).

124 usually ineffective for reducing punishment: Kenneth J. Weiss, Alisa R. Gutman, and Wade H. Berrettini, "Invoking Behavioral Genetics in Criminal Mitigation: What Can Experts Reasonably Say?," SSRN Scholarly Paper, Social Science Research Network, Rochester, N.Y., February 21, 2020, papers.ssrn.com/abstract=3800216.

124 what *is* considered potentially mitigating information: Russell Stetler, "The History of Mitigation in Death Penalty Cases," in *Social Work, Criminal Justice, and the Death Penalty,* ed. Lauren A. Ricciardelli (New York: Oxford University Press, 2020), doi.org/10.1093/oso/9780190937232.003.0004; Laurence Steinberg, "The Influence of Neuroscience on US Supreme Court Decisions About Adolescents' Criminal Culpability," *Nature Reviews Neuroscience* 14, no. 7 (July 2013): 513–18, doi.org/10.1038/nrn3509.

125 might have been shaped by trauma: Deborah W. Denno, "How Courts in Criminal Cases Respond to Childhood Trauma," SSRN Scholarly

Paper, Social Science Research Network, Rochester, N.Y., February 19, 2020, papers.ssrn.com/abstract=3541194.

126 **motivated to (re)invent a notion of will**: Michael Frede, *A Free Will: Origins of the Notion in Ancient Thought* (Berkeley: University of California Press, 2011), jstor.org/stable/10.1525/j.ctt1ppdd9.

126 **challenge posed to the church by Gnosticism**: Pagels, *Adam, Eve, and the Serpent;* Frede, *A Free Will*.

126 **"For Augustine, natural and moral evils"**: Pagels, *Adam, Eve, and the Serpent*.

127 **"The body keeps the score"**: Bessel van der Kolk, *The Body Keeps the Score: Brain, Mind, and Body in the Healing of Trauma* (New York: Penguin, 2015).

127 **"Whatever is natural"**: Pagels, *Adam, Eve, and the Serpent*, 142.

128 **"hope of humanity's capacity"**: Pagels, *Adam, Eve, and the Serpent*, 149.

131 **"We are all broken"**: Bryan Stevenson, *Just Mercy: A Story of Justice and Redemption* (New York: One World, 2015).

132 **the sociologist Max Weber described**: Max Weber, *The Protestant Ethic and the Spirit of Capitalism*, trans. Talcott Parsons (Kettering, Ohio: Angelico Press, 2014).

133 **"conduct shows irretrievable depravity"**: Juvenile Sentencing Project, "Irreparable Corruption," accessed March 17, 2025, juvenilesentencingproject.org/category/significant-case-law/corruption.

133 **an essence placeholder**: Ilan Dar-Nimrod and Steven J. Heine, "Genetic Essentialism: On the Deceptive Determinism of DNA," *Psychological Bulletin* 137, no. 5 (September 2011): 800–818, doi.org/10.1037/a0021860.

133 **"Sheep and goats have different DNA"**: Hayden Hefner, "The Real Difference Between Sheep and Goats," *The Gospel Coalition*, July 15, 2021, thegospelcoalition.org/article/real-difference-sheep-goats.

134 **philosopher Gary Watson discusses**: Watson, "Responsibility and the Limits of Evil."

137 **"Essentialist biases allow"**: Dar-Nimrod and Heine, "Genetic Essentialism."

137 **"Here grows the Cure"**: Milton, *Paradise Lost*.

Variety

140 **"by what logic"**: Louise Glück, "Clover," *The Wild Iris* (Hopewell, N.J.: Ecco, 1993).

141 **"The Garden was prolific"**: Emilia Phillips, "Book X: The Queerness of Eve," August 19, 2021, theadroitjournal.org/issue-thirty-eight/emilia-phillips.

144 **"substantial impairment"**: "Human Life Protection Act (H.B. 1280)" (2021), capitol.texas.gov/tlodocs/87R/billtext/html/HB01280I.htm.

144 **Americans did not have**: *Dobbs v. Jackson Women's Health Organization*, 597 U.S. 215 (U.S. Supreme Court, 2022).

144 **ten-year-old Ohio girl**: David Folkenflik and Sarah McCammon, "A Rape, an Abortion, and a One-Source Story: A Child's Ordeal Becomes

National News," NPR, July 13, 2022, npr.org/2022/07/13/1111285143/abortion-10-year-old-raped-ohio.
- 145 "exhausted by a married life": Shannon K. Withycombe, "From Women's Expectations to Scientific Specimens: The Fate of Miscarriage Materials in Nineteenth-Century America," *Social History of Medicine* 28, no. 2 (May 1, 2015): 245–62, doi.org/10.1093/shm/hku071.
- 147 "We think some infants": Helga Kuhse and Peter Singer, *Should the Baby Live? The Problem of Handicapped Infants* (Oxford: Oxford University Press, 1988).
- 147 "cannot see themselves as beings": Peter Singer, *Practical Ethics*, 2nd ed. (Cambridge: Cambridge University Press, 1993), 131, 171.
- 147 "the man who wants me dead": Harriet McBryde Johnson, "Unspeakable Conversations," *The New York Times Magazine,* February 16, 2003, nytimes.com/2003/02/16/magazine/unspeakable-conversations.html.
- 147 "life unworthy of life": Howard Brody and M. Wayne Cooper, "Binding and Hoche's 'Life Unworthy of Life': A Historical and Ethical Analysis," *Perspectives in Biology and Medicine* 57, no. 4 (2014): 500–511, doi.org/10.1353/pbm.2014.0042.
- 147 "eugenics dragnet widened": Harriet McBryde Johnson, "Wheelchair Unbound," *The New York Times Magazine,* April 23, 2006, nytimes.com/2006/04/23/magazine/wheelchair-unbound.html.
- 148 The Nazi "Law for the Prevention of Offspring": Kevles, *In the Name of Eugenics.*
- 148 "It is better for all": *Buck v. Bell,* 274 U.S. 200 (U.S. Supreme Court, 1927).
- 148 The word *eugenics:* Adam Rutherford, *Control: The Dark History and Troubling Present of Eugenics* (New York: W. W. Norton, 2022).
- 149 Davenport published an account: Charles Benedict Davenport, *The Study of Human Heredity: Methods of Collecting, Charting, and Analyzing Data* (Cold Spring Harbor, N.Y.: Eugenics Record Office, 1911), archive.org/details/studyofhumanhere02dave.
- 149 who sought alliances: Dorothy Roberts, "Margaret Sanger and the Racial Origins of the Birth Control Movement," in *Racially Writing the Republic: Racists, Race Rebels, and Transformations of American Identity,* ed. Bruce Baum and Duchess Harris (Durham, N.C.: Duke University Press, 2009), 196–214, doi.org/10.1515/9780822392156-012.
- 149 "unite the Eugenic Movement": Jonathan Spiro, *Defending the Master Race: Conservation, Eugenics, and the Legacy of Madison Grant* (Burlington: University of Vermont Press, 2009).
- 149 administer IQ tests to immigrants: Kevles, *In the Name of Eugenics;* Rutherford, *Control.*
- 149 In Goddard's subsequent book: Henry Herbert Goddard, *The Kallikak Family: A Study in the Heredity of Feeble-Mindedness* (New York: Macmillan Co., 1912), doi.org/10.1037/10949-000.
- 150 his purported pedigrees: Rutherford, *Control.*
- 150 Ruth Whitfield, who was survived: Kiara Alfonseca, "Funerals for Buffalo Shooting Victims Continue as Nation Grapples with Gun Violence,"

Good Morning America, accessed March 17, 2025, goodmorningamerica.com/news/story/funerals-buffalo-shooting-victims-continue-nation-grapples-gun-85016980.

151 **"terrorize all nonwhite"**: "Buffalo Shooting Suspect's Prior Threat Sent Him to Mental Hospital," May 16, 2022, cbsnews.com/news/buffalo-shooting-suspects-prior-threat-mental-hospital.

151 **middle school students in the United States**: Brian M. Donovan et al., "Toward a More Humane Genetics Education: Learning About the Social and Quantitative Complexities of Human Genetic Variation Research Could Reduce Racial Bias in Adolescent and Adult Populations," *Science Education* 103, no. 3 (2019): 529–60, doi.org/10.1002/sce.21506.

151 **nearly all genetic variation exists**: Rutherford, *How to Argue with a Racist*.

151 **Racial categories are constructed**: Michael Yudell et al., "Taking Race out of Human Genetics," *Science* 351, no. 6273 (February 5, 2016): 564–65, doi.org/10.1126/science.aac4951.

151 **they show less racial bias**: Donovan et al., "Toward a More Humane Genetics Education."

151 **They are, for instance, less likely**: Rutherford, *How to Argue with a Racist*.

151 **Could Gendron have been inoculated**: Kathryn Paige Harden, "A Book That Could Save Lives: Adam Rutherford's *How to Argue with a Racist* Reviewed," *The Spectator*, March 14, 2020, spectator.co.uk/article/a-book-that-could-save-lives-adam-rutherford-s-how-to-argue-with-a-racist-reviewed.

152 **Could he and others be turned**: Jedidiah Carlson et al., "Counter the Weaponization of Genetics Research by Extremists," *Nature* 610, no. 7932 (October 2022): 444–47, doi.org/10.1038/d41586-022-03252-z.

152 **"There can be no mercy"**: Bill Hutchinson and Aaron Katersky, "Judge Says 'No Mercy,' Victims' Families Vent Their Anger as Buffalo Mass Shooter Sentenced to Life," ABC News, February 15, 2023, abcnews.go.com/US/buffalo-mass-shooter-payton-gendron-set-sentenced-life/story?id=97079974.

152 **The 2021 case before the Sixth Circuit**: *Preterm-Cleveland v. McCloud*, 994 F.3d. 512 (6th Cir. 2021).

152 **between 60 and 90 percent of pregnancies**: Joint Economic Committee Republicans, "Down Syndrome and Social Capital: Assessing the Costs of Selective Abortion," Social Capital Project, United States Congress Joint Economic Committee—Republicans, Washington, D.C., March 2022) jec.senate.gov/public/index.cfm/republicans/2022/3/down-syndrome-and-social-capital-assessing-the-costs-of-selective-abortion.

153 **"refute the '*it*'-ness"**: Michael Stokes Paulson, "Abortion as an Instrument of Eugenics," *Harvard Law Review* 134, no. 7 (May 2021): 415–33.

153 **language comparing abortion to eugenics**: Reva B. Siegal and Mary Ziegler, "Abortion-Eugenics Discourse in Dobbs: A Social Movement History," *Journal of American Constitutional History*, February 14, 2024, jach.law.wisc.edu/abortion-eugenics-discourse-in-dobbs.

153 **seventh child has Down syndrome**: Margaret Talbot, "Amy Coney Bar-

rett's Long Game," *The New Yorker,* February 7, 2022, newyorker.com/magazine/2022/02/14/amy-coney-barretts-long-game.

153 **"There is a difference between"**: *Planned Parenthood of Ind. & Ky., Inc. v. Comm'r of the Ind. State Dep't of Health,* 917 F.3d 532 (7th Cir. 2018).

154 **cited a 2018 *New York Times* article**: Clyde Haberman, "Scientists Can Design 'Better' Babies. Should They?," *The New York Times,* June 11, 2018, nytimes.com/2018/06/10/us/11retro-baby-genetics.html.

154 **The first human child**: Kalina Kamenova and Hazar Haidar, "The First Baby Born After Polygenic Embryo Screening: Key Issues Through the Lens of Experts and Science Reporters," *Voices in Bioethics* 8 (April 7, 2022), doi.org/10.52214/vib.v8i.9467.

155 **a podcast interview with Steve Hsu**: Steve Hsu, "Simone Collins: IVF, Embryo Selection, Dating on the Spectrum, and Pronatalism," *Manifold* (blog), no. 34, April 27, 2023, share.transistor.fm/s/2c01ce75.

155 **"amoral nonsense"**: Lior Pachter, "The Amoral Nonsense of Orchid's Embryo Selection," *Bits of DNA* (blog), April 12, 2021, liorpachter.wordpress.com/2021/04/12/the-amoral-nonsense-of-orchids-embryo-selection.

155 **"womb goblins"**: Diana Fleischman, Ives Parr, and Laurent Tellier, "Embryo Selection: Toward a Healthier Society," *Aporia,* April 6, 2023, aporiamagazine.com/p/embryo-selection-healthy-babies-vs.

156 **"preimplantation genetic testing"**: Aya Abu-El-Haija et al., "The Clinical Application of Polygenic Risk Scores: A Points to Consider Statement of the American College of Medical Genetics and Genomics (ACMG)," *Genetics in Medicine* 25, no. 5 (May 1, 2023), doi.org/10.1016/j.gim.2023.100803.

156 **"the Pandora's box of embryo testing"**: Carey Goldberg, "The Pandora's Box of Embryo Testing Is Officially Open," *Bloomberg,* May 26, 2022, bloomberg.com/news/features/2022-05-26/dna-testing-for-embryos-promises-to-predict-genetic-diseases.

156 **A majority (58 percent)**: Michelle N. Meyer et al., "Public Views on Polygenic Screening of Embryos," *Science* 379, no. 6632 (February 10, 2023): 541–43, doi.org/10.1126/science.ade1083.

156 **"pits unborn siblings"**: Jordan Boyd, "Popular Genetic Testing Pits Unborn Siblings Against Each Other, and the Highest Scorer Gets to Live," *The Federalist,* June 8, 2023, thefederalist.com/2023/06/08/popular-genetic-testing-pits-unborn-siblings-against-each-other-and-the-highest-scorer-gets-to-live.

157 **"natural limits upon"**: Albert Mohler, "The Briefing," September 21, 2021, albertmohler.com/2021/09/21/briefing-9-21-21.

159 *expected breeding values:* Naomi R. Wray et al., "Complex Trait Prediction from Genome Data: Contrasting EBV in Livestock to PRS in Humans: Genomic Prediction," *Genetics* 211, no. 4 (April 1, 2019): 1131–41, doi.org/10.1534/genetics.119.301859.

159 **Consider the chicken**: M. J. Zuidhof et al., "Growth, Efficiency, and Yield of Commercial Broilers from 1957, 1978, and 2005," *Poultry Science* 93, no. 12 (December 1, 2014): 2970–82, doi.org/10.3382/ps.2014-04291.

160 "there seems to be a limit": Susan Wolf, "Moral Saints," *The Journal of Philosophy* 79, no. 8 (1982): 419–39, doi.org/10.2307/2026228.
161 we are of the Devil's party: Blake, *Marriage of Heaven and Hell*.
161 one of the ways my colleagues and I: Karlsson Linnér et al., "Multivariate Analysis of 1.5 Million People Identifies Genetic Associations with Traits Related to Self-Regulation and Addiction."
161 "You should know": Glück, "Clover," *Wild Iris*.
161 evolutionary forces have maintained: Katherine M. Siewert and Benjamin F. Voight, "Detecting Long-Term Balancing Selection Using Allele Frequency Correlation," *Molecular Biology and Evolution* 34, no. 11 (November 2017): 2996–3005, doi.org/10.1093/molbev/msx209.
161 "a factor in public health": Emile Durkheim, *The Rules of Sociological Method*, trans. W. D. Halls (New York: The Free Press, 1982), 98.
162 "But so that the originality": Durkheim, *Rules of Sociological Method*, 101.
162 "If you want to understand": Zach Weisberg, "Yvon Chouinard Explains How He Built Patagonia," *The Inertia* (blog), July 10, 2017, theinertia.com/surf/yvon-chouinard-explains-how-he-built-patagonia.
163 who engaged in antisocial behavior as teenagers: Ross Levine and Yona Rubinstein, "Smart and Illicit: Who Becomes an Entrepreneur and Do They Earn More?," *Quarterly Journal of Economics* 132 (May 1, 2017): 963–1018, doi.org/10.1093/qje/qjw044.
163 We call these cheating cells "cancer": Athena Aktipis, *The Cheating Cell: How Evolution Helps Us Understand and Treat Cancer* (Princeton: Princeton University Press, 2020).
163 cheating can become the new normal: Athena Aktipis et al., "Cancer Across the Tree of Life: Cooperation and Cheating in Multicellularity," *Philosophical Transactions of the Royal Society B: Biological Sciences*, July 19, 2015, doi.org/10.1098/rstb.2014.0219.

Consequences

167 one of the bestselling board books: Elizabeth Verdick, *Teeth Are Not for Biting* (Minneapolis: Free Spirit Publishing, 2003).
167 Crying is contagious: Abraham Sagi and Martin L. Hoffman, "Empathic Distress in the Newborn," *Developmental Psychology* 12, no. 2 (1976): 175–76, doi.org/10.1037/0012-1649.12.2.175; Elena Geangu et al., "Contagious Crying Beyond the First Days of Life," *Infant Behavior and Development* 33, no. 3 (June 1, 2010): 279–88, doi.org/10.1016/j.infbeh.2010.03.004; Grace B. Martin and Russell D. Clark, "Distress Crying in Neonates: Species and Peer Specificity," *Developmental Psychology* 18, no. 1 (1982): 3–9, doi.org/10.1037/0012-1649.18.1.3; Paul Bloom, *Just Babies: The Origins of Good and Evil* (New York: Crown, 2014).
168 their emotional empathy for other people's distress: Paul J. Frick and Emily C. Kemp, "Conduct Disorders and Empathy Development," *Annual Review of Clinical Psychology* 17 (May 7, 2021): 391–416, doi.org/10.1146/annurev-clinpsy-081219-105809.
168 activates multiple areas of the brain: Jamil Zaki and Kevin N. Ochsner,

"The Neuroscience of Empathy: Progress, Pitfalls and Promise," *Nature Neuroscience* 15, no. 5 (May 2012): 675–80, doi.org/10.1038/nn.3085; Frans B. M. de Waal and Stephanie D. Preston, "Mammalian Empathy: Behavioural Manifestations and Neural Basis," *Nature Reviews Neuroscience* 18, no. 8 (August 2017): 498–509, doi.org/10.1038/nrn.2017.72.

168 **not all children experience affective empathy:** Frick and Kemp, "Conduct Disorders and Empathy Development."

169 **"It is the emotionally charged":** Robert D. Hare, *Without Conscience: The Disturbing World of the Psychopaths Among Us* (New York: Pocket Books, 1995), 132.

169 **Lower levels of cognitive:** Carlos Campos et al., "Refining the Link Between Psychopathy, Antisocial Behavior, and Empathy: A Meta-Analytical Approach," *Clinical Psychology Review* 94 (2022): 1021–45.

170 **"Law of Effect":** Edward L. Thorndike, "The Law of Effect," *American Journal of Psychology* 39, no. 1/4 (1927): 212–22, doi.org/10.2307/1415413.

170 **trigger the release of dopamine:** Ewa Galaj and Robert Ranaldi, "Neurobiology of Reward-Related Learning," *Neuroscience and Biobehavioral Reviews* 124 (May 2021): 224–34, doi.org/10.1016/j.neubiorev.2021.02.007.

170 **"unpleasant sensory and emotional experience":** Srinivasa N. Raja et al., "The Revised International Association for the Study of Pain Definition of Pain: Concepts, Challenges, and Compromises," *Pain* 161, no. 9 (September 1, 2020): 1976–82, doi.org/10.1097/j.pain.0000000000001939.

171 **nociception, the ability to sense injury:** Edgar T. Walters and Amanda C. de C. Williams, "Evolution of Mechanisms and Behaviour Important for Pain," *Philosophical Transactions of the Royal Society B: Biological Sciences* 374, no. 1785 (November 11, 2019): 20190275, doi.org/10.1098/rstb.2019.0275.

171 **Consider the lobster:** David Foster Wallace, *Consider the Lobster and Other Essays* (New York: Back Bay Books, 2007).

171 **eat the larvae alive:** Jessica L. Robertson, Asako Tsubouchi, and W. Daniel Tracey, "Larval Defense Against Attack from Parasitoid Wasps Requires Nociceptive Neurons," *PLoS ONE* 8, no. 10 (October 25, 2013): e78704, doi.org/10.1371/journal.pone.0078704.

171 **Children who have mutations:** Khadije Daneshjou, Hanieh Jafarieh, and Seyed-Reza Raaeskarami, "Congenital Insensitivity to Pain and Anhydrosis (CIPA) Syndrome: A Report of 4 Cases," *Iranian Journal of Pediatrics* 22, no. 3 (September 2012): 412.

171 **"For Zeus' law is first":** Chappell, "Agamemnon at Aulis."

172 **Skinner's data had made him suspect:** Bruna Colombo dos Santos and Marcus Bentes de Carvalho Neto, "B. F. Skinner's Evolving Views of Punishment: II. 1940–1960," *Revista Mexicana de Análisis de La Conducta* 46, no. 2 (2020): 293–318.

172 **"The old school made":** B. F. Skinner, *Walden Two* (Indianapolis: Hackett, 2005).

173 **Luke Hyde and his colleagues studied:** Luke W. Hyde et al., "Heritable and Nonheritable Pathways to Early Callous-Unemotional Behaviors,"

American Journal of Psychiatry 173, no. 9 (September 2016): 903–10, doi.org/10.1176/appi.ajp.2016.15111381.

174 **one of the most effective treatments:** Alan E. Kazdin et al., "Parent Management Training for Conduct Problems in Children: Enhancing Treatment to Improve Therapeutic Change," *International Journal of Clinical and Health Psychology* 18, no. 2 (2018): 91–101, doi.org/10.1016/j.ijchp.2017.12.002.

176 **50 percent of U.S. parents:** Christopher J. Mehus and Megan E. Patrick, "Prevalence of Spanking in US National Samples of 35-Year-Old Parents from 1993 to 2017," *JAMA Pediatrics* 175, no. 1 (January 1, 2021): 92–94, doi.org/10.1001/jamapediatrics.2020.2197.

176 **only 35 percent of U.S. parents:** Mehus and Patrick, "Prevalence of Spanking in US National Samples of 35-Year-Old Parents from 1993 to 2017."

176 **only 20 percent of ten-year-olds:** "One in Five 10-Year-Olds Experience Physical Punishment," *UCL News,* May 1, 2024, ucl.ac.uk/news/2024/may/one-five-10-year-olds-experience-physical-punishment.

176 **74 percent of millennials:** "Millennials Like to Spank Their Kids Just as Much as Their Parents Did," *The Washington Post,* March 5, 2015, accessed February 28, 2025, washingtonpost.com/news/wonk/wp/2015/03/05/millennials-like-to-spank-their-kids-just-as-much-as-their-parents-did.

176 **more than 71 percent of English adults:** "One in Five 10-Year-Olds Experience Physical Punishment."

176 **"clear and unconditional prohibition":** "Forms of Punishment," U.N. Special Representative of the Secretary-General on Violence Against Children, violenceagainstchildren.un.org/content/forms-punishment; Hannah Lichtsinn and Jeffrey Goldhagen, "Why the USA Should Ratify the UN Convention on the Rights of the Child," *BMJ Paediatrics Open* 7, no. 1 (February 2023), doi:10.1136/bmjpo-2021-001355.

176 **One paper surveyed the results:** Elizabeth T. Gershoff and Andrew Grogan-Kaylor, "Spanking and Child Outcomes: Old Controversies and New Meta-Analyses," *Journal of Family Psychology* 30, no. 4 (June 2016): 453–69, doi.org/10.1037/fam0000191.

177 **eleven-year-old children were put:** Jorge Cuartas et al., "Corporal Punishment and Elevated Neural Response to Threat in Children," *Child Development* 92, no. 3 (May 2021): 821–32, doi.org/10.1111/cdev.13565.

177 **Support for spanking:** Harry Enten, "Americans' Opinions on Spanking Vary by Party, Race, Region, and Religion," *FiveThirtyEight* (blog), September 15, 2014, fivethirtyeight.com/features/americans-opinions-on-spanking-vary-by-party-race-region-and-religion.

178 **Enthusiasm among U.S. Christians:** Anne-Marie Cusac, "Flogging for Jesus," *Cruel and Unusual: The Culture of Punishment in America* (New Haven: Yale University Press, 2009), 134–38; Philip Greven, *Spare the Child: The Religious Roots of Punishment and the Psychological Impact of Abuse* (New York: Alfred A. Knopf, 1991).

178 **Esther Edwards Burr:** Cusac, *Cruel and Unusual,* 136.

178 **an array of books:** Talia Levin, *Wild Faith: How the Christian Right Is*

Taking Over America (New York: Legacy Lit, 2024); Alice Miller, *For Your Own Good: Hidden Cruelty in Child-Rearing and the Roots of Violence* (Farrar, Straus and Giroux, 2002); Cusac, *Cruel and Unusual*.

178 **Shepherding:** Tedd Tripp, *Shepherding a Child's Heart* (Wapwallopen, Penn.: Shepherd Press, 2011); J. Richard Fugate and Daniel Carlen, *What the Bible Says About Child Training: Parenting with Confidence*, ed. Dr. Myron Yeager (Apache Junction, Ariz.: Foundation for Biblical Research, 2013); Roy Lessin, *Spanking: A Loving Discipline: Helpful and Practical Answers for Today's Parents* (Ada, Mich.: Bethany House, 2002); James C. Dobson, *The New Strong-Willed Child* (Carol Stream, Ill.: Tyndale Momentum, 2014); Larry Tomczak, *God, the Rod, and Your Child's Bod: The Art of Loving Correction for Christian Parents* (Old Tappan, N.J.: Fleming H. Revell, 1982).

178 **"To blame sin"**: Ginger Hubbard, *Don't Make Me Count to Three: A Mom's Look at Heart-Oriented Discipline* (Wapwallopen, Penn.: Shepherd Press, 2004).

178 ***To Train Up a Child:*** Michael Pearl and Debi Pearl, *To Train Up a Child* (Pleasantville: No Greater Joy Ministries, 1994).

178 **circumstantially linked to the deaths:** "Child 'Training' Book Triggers Backlash," BBC News, December 11, 2013, bbc.com/news/magazine-25268343.

179 **"To give up the rod":** Michael Pearl, "In Defense of Biblical Chastisement, Part 1," *No Greater Joy Ministries* (blog), April 15, 2001, nogreaterjoy.org/articles/in-defense-of-biblical-chastisement-part-1.

179 ***Good Inside:*** Becky Kennedy, *Good Inside: A Guide to Becoming the Parent You Want to Be* (New York: Harper Wave, 2022).

179 **Spanking by the person:** Miller, *For Your Own Good*.

180 **"You act like the world":** Noah Hawley, *Anthem* (New York: Grand Central Publishing, 2022).

180 **"Lawlessness is more likely to occur":** James C. Dobson, *Dare to Discipline* (Wheaton: Living Books, 1987).

181 **"a religious history of mass incarceration":** Aaron Griffith, *God's Law and Order* (Cambridge: Harvard University Press, 2020), doi.org/10.4159/9780674249776.

181 ***Dare to Discipline*** **illustrated one way:** The same connection between political authoritarianism and support for corporal punishment can be seen today. For instance, in 2024, the right-wing political commentator Tucker Carlson approvingly compared then–presidential candidate Donald Trump to a punitive father: "Dad comes home. He's pissed. Dad is pissed. And when dad gets home, you know what he says? 'You've been a bad girl. You've been a bad little girl, and you're getting a vigorous spanking right now. I'm not going to lie. It's going to hurt you a lot more than it hurts me. And you earned this. You're getting a vigorous spanking because you've been a bad girl. You're only going to get better when you take responsibility for what you did. It has to be this way.'" Arwa Mahdawi, "Tucker Carlson Is Fantasizing About Daddy Donald Trump Spanking Teenage Girls," *The Guardian*, October 24, 2024, theguardian.com/us-news/2024/oct/24/tucker-carlson-speech-trump-spanking.

- 181 **161 per 100,000 people:** "World Prison Brief," Institute for Crime and Justice Policy Research, accessed March 17, 2025, prisonstudies.org/country/united-states-america.
- 181 **In 1971, President Nixon:** Ashley Nellis, "Mass Incarceration Trends," The Sentencing Project, May 21, 2024, sentencingproject.org/reports/mass-incarceration-trends.
- 181 **Republicans and Democrats competed:** Lauren-Brooke Eisen, "The 1994 Crime Bill and Beyond: How Federal Funding Shapes the Criminal Justice System," Brennan Center for Justice, September 9, 2019, brennancenter.org/our-work/analysis-opinion/1994-crime-bill-and-beyond-how-federal-funding-shapes-criminal-justice.
- 181 **More money was poured:** Michelle Alexander, *The New Jim Crow: Mass Incarceration in the Age of Colorblindness* (New York: The New Press, 2012).
- 181 **755 per 100,000:** "World Prison Brief."
- 181 **For comparison, the U.K.:** Prison Policy Initiative, "States of Incarceration: The Global Context 2024," accessed February 28, 2025, prisonpolicy.org/global/2024.html.
- 182 **punishing people more harshly:** Gary S. Becker, "Crime and Punishment: An Economic Approach," *The Journal of Political Economy* 76, no. 2 (1968): 169–217.
- 182 **the death penalty deterred:** Isaac Ehrlich, "The Deterrent Effect of Capital Punishment: A Question of Life and Death," *American Economic Review* 65, no. 3 (1975): 397–417.
- 182 **the U.S. Supreme Court ruled:** *Gregg v. Georgia*, 428 U.S. 153 (1976).
- 182 **The death penalty is outlawed:** Seren Morris, Dan Cody, and Ayan Omar, "When Did the Death Penalty Become Illegal in the UK? Alabama Killer Was First to Be Executed with Nitrogen Gas," *The Standard*, January 26, 2024, standard.co.uk/news/uk/death-penalty-law-illegal-uk-world-alabama-nitrogen-gas-b1059110.html.
- 182 **"the infliction of death":** *Gregg v. Georgia*, 428 U.S. 153 (1976).
- 183 **"broken windows" theory of policing:** James Q. Wilson, *Thinking About Crime* (New York: Basic Books, 1975).
- 183 **"tough-on-crime" policies:** Wilson, *Thinking About Crime*, 117.
- 183 *Crime and Human Nature:* James Q. Wilson and Richard J. Herrnstein, *Crime and Human Nature: The Definitive Study of the Causes of Crime* (New York: Free Press, 1998).
- 183 **Richard Herrnstein:** Richard J. Herrnstein, *I.Q. in the Meritocracy* (New York: Little, Brown, 1973); Herrnstein and Murray, *The Bell Curve*.
- 183 **"Wicked people exist":** Wilson, *Thinking About Crime*, 260.
- 185 **recommended the abolition of parole:** Michael Wright and Caroline Rand Herron, "The Nation in Summary: The Hard Line Against Crime," *The New York Times*, July 26, 1981, nytimes.com/1981/07/26/weekinreview/the-nation-in-summary-the-hard-line-against-crime.html; U.S. Attorney General, "Attorney General's Task Force on Violent Crime," public hearing, Key Biscayne, Florida, U.S. Department of Justice, Office of Justice Programs, 1981, ojp.gov/ncjrs/virtual-library/abstracts/attorney-generals-task-force-violent-crime-public-hearing-key.

- 185 Justice Antonin Scalia's line of questioning: *Miller v. Alabama*, 567 U.S. 460 (2012).
- 185 "Well, I thought that": Cited in Blake, *Uninnocent*.
- 185 The United States remains: "Juvenile Life Without Parole (JLWOP)," Juvenile Law Center, accessed March 6, 2025, jlc.org/issues/juvenile-life-without-parole.
- 185 the legacy of the more-punishment: Alex Tabarrok, "What Was Gary Becker's Biggest Mistake?," *Marginal Revolution* (blog), September 16, 2015, marginalrevolution.com/marginalrevolution/2015/09/what-was-gary-beckers-biggest-mistake.html.
- 185 While incarceration rates: Alexander, *New Jim Crow;* Michael Tonry, "Why Crime Rates Are Falling Throughout the Western World," *Crime and Justice* 43 (September 2014): 1–63, doi.org/10.1086/678181.
- 186 As Michelle Alexander described: Alexander, *New Jim Crow*.
- 186 "We are gradually discovering": Skinner, *Walden Two*.
- 187 In his book *When Brute Force Fails:* Mark Kleiman, *When Brute Force Fails: How to Have Less Crime and More Punishment* (Princeton: Princeton University Press, 2010).
- 187 Jennifer Doleac: Jennifer L. Doleac, "Encouraging Desistance from Crime," *Journal of Economic Literature* 61, no. 2 (June 2023): 383–427, doi.org/10.1257/jel.20211536.
- 188 "responsibility without blame": Hanna Pickard, "Responsibility Without Blame: Philosophical Reflections on Clinical Practice," *The Oxford Handbook of Philosophy and Psychiatry*, ed. K.W.M. Fulford et al. (Oxford: Oxford University Press, 2013), 1134–52.
- 188 "The first step": Kleiman, *When Brute Force Fails*, 1, emphasis added.

Retribution

- 190 wounded or vulnerable: Elizabeth A. Tibbetts, Ellery Wong, and Sarah Bonello, "Wasps Use Social Eavesdropping to Learn About Individual Rivals," *Current Biology* 30, no. 15 (August 2020): 3007–10, doi.org/10.1016/j.cub.2020.05.053.
- 190 One-third of all of their children: Hudson K. Reeve and Peter Nonacs, "Social Contracts in Wasp Societies," *Nature* 359, no. 6398 (October 1992): 823–25, doi.org/10.1038/359823a0.
- 192 Worker bees attack: T. H. Clutton-Brock and G. A. Parker, "Punishment in Animal Societies," *Nature* 373, no. 6511 (January 19, 1995): 209–16, http://dx.doi.org/10.1038/373209a0.
- 192 "reactive attitudes": P. F. Strawson, *Freedom and Resentment and Other Essays* (London and New York: Routledge, 2008).
- 192 emergence of the first multicellular organisms: J. Arvid Ågren, Nicholas G. Davies, and Kevin R. Foster, "Enforcement Is Central to the Evolution of Cooperation," *Nature Ecology and Evolution* 3, no. 7 (July 2019): 1018–29, doi.org/10.1038/s41559-019-0907-1.
- 192 there are also cancer cells: Aktipis, *Cheating Cell*.
- 192 "jumping genes": L. E. Orgel and F.H.C. Crick, "Selfish DNA: The Ultimate Parasite," *Nature* 284 (April 1980): 604-7; Richard N. McLaughlin,

Jr. and Harmit S. Malik, "Genetic Conflicts: The Usual Suspects and Beyond," *Journal of Experimental Biology* 220, no. 1 (January 2017): 6–17.

193 **chemistry somehow gave rise to biology:** Cole Mathis et al., "Prebiotic RNA Network Formation: A Taxonomy of Molecular Cooperation," *Life* 7, no. 4 (December 2017): 38, doi.org/10.3390/life7040038.

193 **"is organized in a hierarchy":** Ågren, Davies, and Foster, "Enforcement Is Central to the Evolution of Cooperation."

193 **Evolutionary theories propose:** Richard Alexander, *The Biology of Moral Systems* (New York: Routledge, 2017), doi.org/10.4324/9780203700976; Christopher Boehm, *Moral Origins: The Evolution of Virtue, Altruism, and Shame* (New York: Basic Books, 2012); Léo Fitouchi, Jean-Baptiste André, and Nicolas Baumard, "Moral Disciplining: The Cognitive and Evolutionary Foundations of Puritanical Morality," *Behavioral and Brain Sciences* 46, no. 1 (2023): 1–71, doi.org/10.1017/S0140525X22002047; Michael E. McCullough, Robert Kurzban, and Benjamin A. Tapak, "Cognitive Systems for Revenge and Forgiveness," *Behavioral and Brain Sciences* 36, no. 1 (February 2013): 1–58; Michael Bang Petersen et al., "To Punish or Repair? Evolutionary Psychology and Lay Intuitions About Modern Criminal Justice," *Evolution and Human Behavior* 33 (2012): 682–95; P. Kyle Stanford, "The Difference Between Ice Cream and Nazis: Moral Externalization and the Evolution of Human Cooperation," *Behavioral and Brain Sciences* 41 (January 2018): e95, doi.org/10.1017/S0140525X17001911.

193 **"Any animal whatever":** Charles Darwin, *The Descent of Man and Selection in Relation to Sex* (New York: D. Appleton, 1909).

193 **"building blocks of moral systems":** Jessica C. Flack and Frans B. M. de Waal, "'Any Animal Whatever': Darwinian Building Blocks of Morality in Monkeys and Apes," *Journal of Consciousness Studies* 7, no. 1–2 (2000): 1–29; Frans B. M. de Waal, *The Bonobo and the Atheist: In Search of Humanism Among the Primates* (New York: W. W. Norton & Company, 2014).

194 **"the possibility of sanctioning":** Özgür Gürerk, Bernd Irlenbusch, and Bettina Rockenbach, "The Competitive Advantage of Sanctioning Institutions," *Science* 312, no. 5770 (April 7, 2006): 108–11, doi.org/10.1126/science.1123633. See also *Genetic and Cultural Evolution of Cooperation*, ed. Peter Hammerstein (Cambridge: MIT Press, 2003); Joseph Henrich and Michael Muthukrishna, "The Origins and Psychology of Human Cooperation," *Annual Review of Psychology* 72 (January 2021): 207–40, doi.org/10.1146/annurev-psych-081920-042106; Justin W. Martin, Sophia Martin, and Katherine McAuliffe, "Third Party Punishment Promotes Fairness in Children," *Developmental Psychology* 57, no. 6 (June 2021): 927–39, doi.org/10.1037/dev0001183.

195 **An estimated one out of six women:** "Scope of the Problem: Statistics," RAINN (Rape, Abuse, and Incest National Network), accessed October 21, 2023, rainn.org/statistics/scope-problem; "Rape, Sexual Assault and Child Sexual Abuse Statistics," Rape Crisis England & Wales, accessed March 8, 2025, rapecrisis.org.uk/get-informed/statistics-sexual-violence.

196 **an empathic neural response:** De Waal and Preston, "Mammalian Empathy"; Zaki and Ochsner, "Neuroscience of Empathy"; Claus Lamm and

Tania Singer, "The Role of Anterior Insular Cortex in Social Emotions," *Brain Structure and Function* 214, no. 5–6 (June 2010): 579–91, doi.org/10.1007/s00429-010-0251-3.

196 **their brains show activity characteristic of pleasure:** Dominique J.-F. de Quervain et al., "The Neural Basis of Altruistic Punishment," *Science (American Association for the Advancement of Science)* 305, no. 5688 (2004): 1254–58, doi.org/10.1126/science.1100735.

197 **Developmental psychologists Paul Bloom and Karen Wynn:** Bloom, *Just Babies;* J. Kiley Hamlin et al., "How Infants and Toddlers React to Antisocial Others," *Proceedings of the National Academy of Sciences* 108, no. 50 (December 13, 2011): 19931–36, doi.org/10.1073/pnas.1110306108.

197 **the pleasure of retribution emerges:** Natacha Mendes et al., "Preschool Children and Chimpanzees Incur Costs to Watch Punishment of Antisocial Others," *Nature Human Behaviour* 2, no. 1 (January 2018): 45–51, doi.org/10.1038/s41562-017-0264-5.

198 **Retribution is an ancient, evolved mechanism:** Stuart A. West et al., "Ten Recent Insights for Our Understanding of Cooperation," *Nature, Ecology, and Evolution* 5, no. 4 (April 2021): 419–30, doi.org/10.1038/s41559-020-01384-x.

198 **Wasps will bite each other:** K. Charlotte Jandér and Edward Allen Herre, "Host Sanctions and Pollinator Cheating in the Fig Tree–Fig Wasp Mutualism," *Proceedings of the Royal Society B: Biological Sciences* 277, no. 1687 (January 13, 2010): 1481–88, doi.org/10.1098/rspb.2009.2157.

199 **"blind to the forces":** Bloom, *Just Babies.*

199 **cruelty is a currency:** Nietzsche, *On the Genealogy of Morals.*

199 **"desire for group conformity":** Danny Osborne et al., "The Psychological Causes and Societal Consequences of Authoritarianism," *Nature Reviews Psychology* 2, no. 4 (April 2023): 220–32, doi.org/10.1038/s44159-023-00161-4.

200 **authoritarianism is as heritable:** Steven G. Ludeke and Robert F. Krueger, "Authoritarianism as a Personality Trait: Evidence from a Longitudinal Behavior Genetic Study," *Personality and Individual Differences* 55, no. 5 (September 1, 2013): 480–84, doi.org/10.1016/j.paid.2013.04.015.

200 **"there could be no forgiveness":** "Punisher (Frank Castle)," accessed March 18, 2025, marvel.com/characters/punisher-frank-castle/in-comics.

201 **Kyle was murdered:** "Eddie Ray Routh Guilty of American Sniper Chris Kyle's Murder," BBC News, February 25, 2015, bbc.com/news/world-us-canada-31617308.

201 **the Punisher skull could be seen everywhere:** Jon Jackson, "Marvel's Punisher Problem," *Newsweek*, March 10, 2021, newsweek.com/marvels-punisher-problem-1574579.

201 **"I hunt the evil":** Katie Way, "NYPD Removes Punisher Poster from Manhattan Sex Offender Check-In Facility," *Hell Gate*, March 14, 2023, hellgatenyc.com/nypd-removes-punisher-poster-from-manhattan-sex-offender-check-in-facility.

201 **"The skull is a reminder":** "Totenkopf," Wikipedia, February 22, 2025, en.wikipedia.org/w/index.php?title=Totenkopf&oldid=1277042381.

201 "Is he a troubled good guy": Jackson, "Marvel's Punisher Problem."
201 **Anthropologists Alan Fiske and Tage Rai argued:** Alan Page Fiske and Tage Shakti Rai, *Virtuous Violence* (Cambridge: Cambridge University Press, 2014).
203 **nearly 30 percent of Americans:** Maddy Savage, "Family Estrangement: Why Adults Are Cutting off Their Parents," BBC, December 1, 2021, bbc.com/worklife/article/20211201-family-estrangement-why-adults-are-cutting-off-their-parents; Anna Russell, "Why So Many People Are Going 'No Contact' with Their Parents," *The New Yorker*, August 30, 2024, newyorker.com/culture/annals-of-inquiry/why-so-many-people-are-going-no-contact-with-their-parents.
203 **Many adults who have gone:** Kathryn Post, "Adults Raised in the 'Christian Parenting Empire' of the '70s–'90s Push Back," *Religion News*, May 30, 2024, religionnews.com/2024/05/30/adults-raised-in-the-christian-parenting-empire-of-the-70s-90s-push-back; Krispin Mayfield, "Chapter 12: Estrangement in Religious Authoritarian Families," Substack newsletter, *Strongwilled* (blog), November 10, 2024, strongwilled.substack.com/p/chapter-12-estrangement-in-religious.
204 "goitered peasants": Sapolsky, *Determined*.
204 **the philosopher Derk Pereboom has argued:** Derk Pereboom, *Free Will, Agency, and Meaning in Life* (Oxford: Oxford University Press, 2014), 182.
204 **"inappropriate" moral emotions:** Seth Shabo, "Where Love and Resentment Meet: Strawson's Intrapersonal Defense of Compatibilism," *Philosophical Review* 121, no. 1 (2012): 95–124.
205 In his essay "Freedom and Resentment": Strawson, *Freedom and Resentment and Other Essays*.
206 **"The existence of some such framework":** Pamela Hieronymi, *Freedom, Resentment, and the Metaphysics of Morals* (Princeton: Princeton University Press, 2020).
206 **Augustine thought that sexual desire:** Augustine, *The Confessions of Saint Augustine*.
206 **Kant worried that sex:** Allen W. Wood, ed., "Sex," in *Kantian Ethics* (Cambridge: Cambridge University Press, 2007), 224–39, doi.org/10.1017/CBO9780511809651.014.
206 **Sexual reproduction emerged:** Sarah P. Otto, "Sexual Reproduction and the Evolution of Sex," *Nature Education* 1, no. 1 (2008): 182.
208 **In 2012, Adam Lanza:** Andrew Solomon, "The Reckoning," *The New Yorker*, March 10, 2014, newyorker.com/magazine/2014/03/17/the-reckoning.
208 **"Chilling Look at Newtown Killer":** Joseph Berger and Marc Santora, "Chilling Look at Newtown Killer, but No 'Why,'" *The New York Times*, November 25, 2013, nytimes.com/2013/11/26/nyregion/sandy-hook-shooting-investigation-ends-with-motive-still-unknown.html.
209 **demonic idol known as gun rights:** Garry Wills, "Our Moloch," *The New York Review of Books*, December 15, 2012, nybooks.com/online/2012/12/15/our-moloch.

209 "I don't want him dead": Lacy M. Johnson, *The Reckonings: Essays* (New York: Scribner, 2018), 11.
210 "Those who wrong others": Jeffrie G. Murphy and Jean Hampton, *Forgiveness and Mercy* (Cambridge: Cambridge University Press, 1998).
210 "brought down": Conor Murray, "Harvey Weinstein Sentenced to Another 16 Years in Prison: His Hollywood Sexual Abuse Crimes, Explained," *Forbes,* February 23, 2023, forbes.com/sites/conormurray/2023/02/23/harvey-weinstein-sentenced-to-another-16-years-in-prison-his-hollywood-sexual-abuse-crimes-explained.
210 "The Court's decision today": The Office of Minnesota Attorney General Keith Ellison, "Court of Appeals Upholds Derek Chauvin's Convictions in Murder of George Floyd," accessed March 18, 2025, ag.state.mn.us/Office/Communications/2023/04/17_Chauvin.asp. Emphasis added.
211 dominant animals are more likely: Clutton-Brock and Parker, "Punishment in Animal Societies."
212 anthropologist Daniel Fessler wrote: Daniel M. T. Fessler, "Toward An Understanding of the Universality of Second Order Emotions," *Beyond Nature or Nurture: Biocultural Approaches to the Emotions,* ed. A. Hinton (New York: Cambridge University Press, 1999), 75–116.
212 "the key to understanding": Jonathan Haidt, "The Moral Emotions," *Handbook of Affective Sciences* 11 (2003): 852–70.
212 nonhuman mammals make facial expressions: Dacher Keltner and Brenda Buswell, "Embarrassment: Its Distinct Form and Appeasement Functions," *Psychological Bulletin* 122, no. 3 (1997): 250–70.
213 "It's as if forgiveness were": Carl Phillips, "Since You Ask," *Reconnaissance: Poems* (Farrar, Straus and Giroux, 2015).
214 Drawing on evolutionary theory: Nicola Lacey and Hanna Pickard, "To Blame or to Forgive? Reconciling Punishment and Forgiveness in Criminal Justice," *Oxford Journal of Legal Studies* 35, no. 4 (2014): 665–96. See also McCullough, Kurzban, and Tapak, "Cognitive Systems for Revenge and Forgiveness"; Petersen et al., "To Punish or Repair? Evolutionary Psychology and Lay Intuitions About Modern Criminal Justice."
214 "From an evolutionary perspective": Lacey and Picard, "To Blame or to Forgive? Reconciling Punishment and Forgiveness in Criminal Justice," 681.

Eclipse

216 Andrea Yates was convinced: "Andrea Yates," *The Crimes That Changed Us* (HBO), accessed May 13, 2024, investigationdiscovery.com/video/the-crimes-that-changed-us-investigation-discovery-atve-us/andrea-yates.
217 she had already been hospitalized four times: "Andrea Yates Fast Facts," CNN, March 25, 2013, cnn.com/2013/03/25/us/andrea-yates-fast-facts/index.html.
218 Up to 8 percent of the population: Lorna Staines et al., "Psychotic Experiences in the General Population, a Review: Definition, Risk Factors,

Outcomes and Interventions," *Psychological Medicine* 52, no. 15 (n.d.): 3297–3308, doi.org/10.1017/S0033291722002550.

219 **whether they consider "voice hearing" abnormal:** T. M. Luhrmann et al., "Differences in Voice-Hearing Experiences of People with Psychosis in the U.S.A., India and Ghana: Interview-Based Study," *British Journal of Psychiatry: The Journal of Mental Science* 206, no. 1 (January 2015): 41–44, doi.org/10.1192/bjp.bp.113.139048.

219 **Ordinary perception:** Anil Seth, *Being You: A New Science of Consciousness* (New York: Dutton, 2021).

219 **the experience of a volitional self:** Seth, *Being You.*

220 **sphexishness:** *Sphexishness* is a term coined by computer scientist Douglas Hofstadter in a 1982 essay about creativity and mechanization. Hofstadter was inspired by an account of behavior in the *Sphex* wasp. Typically, the wasp will paralyze a cricket, take the cricket near its burrow, check to see that the burrow is in good order, drag the cricket into the burrow, and lay eggs alongside it, leaving the preserved cricket as food for its wasp grubs when they hatch. This seemingly planful and complex behavior, however, can be disrupted by a human experimenter, such that the *Sphex* repeats the burrow-checking step countless times and never proceeds with laying eggs—a "shocking revelation of the mechanical underpinnings in a living creature of what looks like quite reflective behavior." Sphexishness, then, is behavior that is repetitive, habitual, mechanical, unproductive, and uncreative—like a stuck record, an agitated herd of cattle, or a "glassy-eyed gambler" in front of a slot machine. Antisphexishness, in contrast, is "nonmechanicalness" that requires conscious "self-watching." Interestingly, Hofstadter suggested that great human art emerged at the boundary of sphexishness and antisphexishness: "It is at the fuzzy boundary where we can no longer quite maintain the self-watching to a high degree of reliability that our own individual styles, characters, begin to emerge to the world." Douglas R. Hofstadter, "Metamagical Themas: Can Inspiration Be Mechanized?," *Scientific American* 247, no. 3 (September 1982): 18–34, jstor.org/stable/10.2307/24966674. For additional discussion of sphexishness in relation to debates about free will, see Daniel C. Dennett, "Please Don't Feed the Bugbears," *Elbow Room: The Varieties of Free Will Worth Wanting* (Cambridge: MIT Press, 2015), 15–37.

222 **One study of low-income men:** Claire E. Ramsay et al., "From Handcuffs to Hallucinations: Prevalence and Psychosocial Correlates of Prior Incarcerations in an Urban, Predominantly African American Sample of Hospitalized Patients with First-Episode Psychosis," *Journal of the American Academy of Psychiatry and the Law* 39, no. 1 (2011): 57–64.

222 **Another study found:** Doris J. James and Lauren E. Glaze, "Mental Health Problems of Prison and Jail Inmates," Office of Justice Programs, U.S. Department of Justice, September 2006, bjs.ojp.gov/library/publications/mental-health-problems-prison-and-jail-inmates.

223 *The miscreant deserves punishment:* Nietzsche, *On the Genealogy of Morals.*

223 *The belief in the volitional self:* Nathan D. Martin, Davide Rigoni, and Kathleen D. Vohs, "Free Will Beliefs Predict Attitudes Toward Unethical

Notes

Behavior and Criminal Punishment," *Proceedings of the National Academy of Sciences* 114, no. 28 (July 11, 2017): 7325–30, doi.org/10.1073/pnas.1702119114; Cory J. Clark et al., "Free to Punish: A Motivated Account of Free Will Belief," *Journal of Personality and Social Psychology* 106, no. 4 (2014): 501–13, doi.org/10.1037/a0035880.

223 "**I will never again**": *The New Oxford Annotated Bible: New Revised Standard Version with the Apocrypha*, ed. Michael D. Coogan (Oxford and New York: Oxford University Press, 2010), 22, emphasis added.

223 "**Never again will I curse**": *Holy Bible: New International Version* (Grand Rapids: Zondervan, 2011). To compare different translations of the same verse, see biblegateway.com/verse/en/Genesis%208%3A21.

223 "**And God saw that the earth**": Genesis 6:13, *New Oxford Annotated Bible*, 19.

224 **If you read the Bible as literature:** Jack Miles, *Christ: A Crisis in the Life of God* (New York: Vintage Books, 2002).

224 "**If as the Bible suggests**": Gold and Appelbaum, "The Inclination to Evil and the Punishment of Crime—from the Bible to Behavioral Genetics."

225 **first oral medication:** Carrie Macmillan, "What to Know About Zurzuvae, the New Pill to Treat Postpartum Depression," *Yale Medicine News*, September 15, 2023, yalemedicine.org/news/postpartum-depression-pill-zurzuvae-zuranolone.

225 **levels drop precipitously:** Paula J. Brunton, John A. Russell, and Jonathan J. Hirst, "Allopregnanolone in the Brain: Protecting Pregnancy and Birth Outcomes," *Progress in Neurobiology* 113 (February 2014): 106–36, doi.org/10.1016/j.pneurobio.2013.08.005.

225 **Allopregnanolone has also been investigated:** Jonathan J. Hirst et al., "Neurosteroid Replacement Approaches for Improving Outcomes After Compromised Pregnancies and Preterm Birth," *Frontiers in Neuroendocrinology* 76 (January 1, 2025): 101169, doi.org/10.1016/j.yfrne.2024.101169; Claire-Marie Vacher et al., "Placental Endocrine Function Shapes Cerebellar Development and Social Behavior," *Nature Neuroscience* 24, no. 10 (October 2021): 1392–1401, doi.org/10.1038/s41593-021-00896-4; Roisin A. Moloney et al., "Zuranolone Therapy Protects Frontal Cortex Neurodevelopment and Improves Behavioral Outcomes After Preterm Birth," *Brain and Behavior* 14, no. 9 (2024): e70009, doi.org/10.1002/brb3.70009.

225 **no medications that are approved:** Mathias Lillig, "Conduct Disorder: Recognition and Management," *American Family Physician* 98, no. 10 (November 15, 2018): 584–92.

225 "**category error**": Marcus W. Feldman and Jessica Riskin, "Why Biology Is Not Destiny," *New York Review of Books* (April 21, 2022), nybooks.com/articles/2022/04/21/why-biology-is-not-destiny-genetic-lottery-kathryn-harden.

226 **the earliest Christians described:** Pagels, *Adam, Eve, and the Serpent*.

226 **As a young child, Anders Breivik:** Åsne Seierstad, *One of Us: The Story of a Massacre in Norway—and Its Aftermath*, trans. Sarah Death (Farrar, Straus and Giroux, 2016).

228 **an astonishing portrait of Breivik:** Seierstad, *One of Us*.

229 **He lives in three prison cells:** Max Fisher, "A Different Justice: Why Anders Breivik Only Got 21 Years for Killing 77 People," *The Atlantic* (blog), August 24, 2012, theatlantic.com/international/archive/2012/08/a-different-justice-why-anders-breivik-only-got-21-years-for-killing-77-people/261532.

229 **the "normality" principle:** "Full Rights Citizens: The Principle of Normality in Norwegian Prisons," *Justice Trends*, July 24, 2018, justice-trends.press/full-rights-citizens-the-principle-of-normality-in-norwegian-prisons.

229 **Norway now spends three times as much:** "World Prison Brief."

229 **incarcerates people at one-tenth of the rate:** Are Høidal and Nina Hanssen, "The History of the Norwegian Correctional System," in *The Norwegian Prison System* (New York: Routledge, 2022).

229 **one of the lowest recidivism rates in the world:** "Recidivism Rates by Country 2025," *World Population Review*, accessed March 17, 2025, worldpopulationreview.com/country-rankings/recidivism-rates-by-country.

229 *not* **an essential feature of democracy:** R. Ingelhart et al., "World Values Survey" (Madrid, 2014).

230 **Norwegians are just about as likely:** Ingelhart et al., "World Values Survey."

230 **only 5 percent of the country:** "Christianity in Norway," Wikipedia, September 27, 2023, en.wikipedia.org/w/index.php?title=Christianity_in_Norway&oldid=1177417350.

230 **more than 60 percent of Americans:** Harriet Sherwood, "Hell, Yes: Younger Britons More Likely to Believe in Damnation, Study Finds," *The Guardian*, May 19, 2023, theguardian.com/world/2023/may/19/younger-britons-more-likely-to-believe-in-eternal-damnation-study-finds; Justin Nortey, Michael Lipka, and Joshua Alvarado, "Few Americans Blame God or Say Faith Has Been Shaken amid Pandemic, Other Tragedies," Pew Research Center, November 23, 2021, pewresearch.org/religion/2021/11/23/views-on-the-afterlife.

231 **"I was a kid":** Ishmael Beah, *Long Way Gone* (New York: Sarah Crichton Books, 2008).

231 **"Laws usually do not have":** Martha Minow, *When Should Law Forgive?* (W. W. Norton & Company, 2019), 41.

231 **In Uganda, ethnographic researchers suggested:** Grace Akello, Annemiek Richters, and Ria Reis, "Reintegration of Former Child Soldiers in Northern Uganda: Coming to Terms with Children's Agency and Accountability," *Intervention: International Journal of Mental Health, Psychosocial Work, and Counselling in Areas of Armed Conflict* 4, no. 3 (2006): 229–43, doi.org/10.1097/WTF.0b013e3280121c00.

231 **studies of former child soldiers in Mozambique:** Minow, *When Should Law Forgive?*, 52; Peter Aldhous, "How Child Soldiers Can Adapt to Life After War," *New Scientist*, May 7, 2008, newscientist.com/article/mg19826553-400-how-child-soldiers-can-adapt-to-life-after-war.

232 **"Being held responsible":** Angela M. Smith, "Responsibility for Attitudes: Activity and Passivity in Mental Life," *Ethics* 115, no. 2 (2005): 269.

233 who deserves rescue and who deserves to suffer: Donika Kelly, "Sanctuary," *The Renunciations: Poems* (Minneapolis: Graywolf Press, 2021), 33.
233 "Neither innocence nor guilt": Minow, *When Should Law Forgive?*, 67.
233 the Gault site: Tyson Bird, "The Gault Site in Central Texas Reveals New Details About the Oldest North Americans," *Texas Highways* (blog), November 23, 2022, texashighways.com/culture/the-gault-site-in-central-texas-reveals-new-details-about-the-oldest-north-americans.
234 "Seeing a partial eclipse": Annie Dillard, *Teaching a Stone to Talk: Expeditions and Encounters* (New York: Harper Collins, 2009).
235 "an emotional response": Michelle N. Shiota, Dacher Keltner, and Amanda Mossman, "The Nature of Awe: Elicitors, Appraisals, and Effects on Self-Concept," *Cognition and Emotion* 21, no. 5 (2007): 944–63, doi.org/10.1080/02699930600923668.
235 a chimpanzee liberated from long captivity: *Emotional Moment as Chimpanzee Sees Open Sky for First Time After 28 Years in Captivity*, 2023, youtube.com/watch?v=EDsTkZ3edbs.
236 a "self-diminishing" and a "self-expanding" emotion: Tonglin Jiang et al., "The Unique Nature and Psychosocial Implications of Awe," *Nature Reviews Psychology* 3, no. 7 (July 2024): 475–88, doi.org/10.1038/s44159-024-00322-z.
236 "Why is she worshipped": St. Augustine, *The City of God*, vol. 2, trans. Marcus Dods (New York: Hafner, 1948), IV.18, 155–56.
237 both Jupiter's mother and Jupiter's child: Daniele Miano, *Fortuna: Deity and Concept in Archaic and Republican Italy* (Oxford: Oxford University Press, 2018), doi.org/10.1093/oso/9780198786566.001.0001.
237 One Hebrew word: Jacob Milgrom, *Leviticus 1–16: A New Translation with Introduction and Commentary*, The Anchor Yale Bible Commentaries (New York: Doubleday, 1991), 1079–84.

Puzzle

239 About 1 in 80,000 apple trees: The Kansas Gardener, "Fruit Tree Guide—Growing Apples," *Grimm's Gardens* (blog), August 12, 2021, grimmsgardens.com/fruit-tree-guide-growing-apples.
240 They evolved to attract: Robert Nicholas Spengler, "Origins of the Apple: The Role of Megafaunal Mutualism in the Domestication of Malus and Rosaceous Trees," *Frontiers in Plant Science* 10 (May 27, 2019): 617, doi.org/10.3389/fpls.2019.00617.
240 The modern apple brings together: Spengler, "Origins of the Apple."
240 The Golden Delicious apple: Barrie Juniper, "The Mysterious Origin of the Sweet Apple," *American Scientist*, February 6, 2017, americanscientist.org/article/the-mysterious-origin-of-the-sweet-apple.
241 The history of the apple transcends: Michael Pollan, *Second Nature: A Gardener's Education* (New York: Grove, 2007).
241 "A good tree cannot bear bad fruit": Matthew 7:18–20 (New King James Version), accessed March 16, 2025, biblehub.com/nkjv/matthew/7.htm.
242 a "bad apple from a bad tree": *Pinholster v. Ayers*, 590 F.3d. 651 (9th Cir. 2009).

242 the "essence placeholder": Dar-Nimrod and Heine, "Genetic Essentialism."
242 "But human excellence": Nussbaum, *Fragility of Goodness*.
243 "confronts us with a deep dilemma": Nussbaum, *Fragility of Goodness*.
245 makes not one whit of difference: Robert Plomin, J. C. DeFries, and John C. Loehlin, "Genotype-Environment Interaction and Correlation in the Analysis of Human Behavior," *Psychological Bulletin* 84, no. 2 (March 1977): 309–22, doi.org/10.1037/0033-2909.84.2.309.
245 "Perhaps we should rephrase": Keller, *Mirage of a Space Between Nature and Nurture*.
246 "druggable" targets: Minikel et al., "Refining the Impact of Genetic Evidence on Clinical Success."
246 "substitution / of the immutable": Louise Glück, *Meadowlands* (Hopewell, N.J.: Ecco, 1997).
248 "There is in these stories": Martha Minow, *Between Vengeance and Forgiveness: Facing History After Genocide and Mass Violence* (Boston: Beacon Press, 1999).

Index

Abel, 40
abortion, 143–44, 145, 152–57
accountability. *See* responsibility
Adam, Eve, and the Serpent (Pagels), 126–27
Adam and Eve
  blaming each other, 109
  curse of God and, 41–42
  God as serpent in Garden, 75
  introduction of evil into world by, 74
  original sin and, 19–20
  punishment of insubordinate flesh, 106–7, 206
  retelling of fable about, 36, 74
  salvation only through God, 21
  shame and, 39
addiction
  behavioral genetics and, 26, 80–81, 257n26
  predicting, using DNA, 16–17
  as scientific term for some sins, 18
  See also drug addiction
"Addiction Is a Brain Disease, and It Matters" (Leshner), 65–66
adoption studies, 49, 80, 114, 173
Aeschylus, 171, 203–4
agency. *See* free will
aggressive behavior. *See* violence
Alexander, Michelle, 186
allopregnanolone, 225
Alypius, 106

American College of Medical Genetics and Genomics, 155–56
American Economic Review, 182
American Psychiatric Association, 76, 78
American Sniper (Kyle), 201
Anastasi, Anne, 119
antisocial behavior(s)
  addiction, 26, 67–68, 80–81, 257n26
  as antidote for blame, 23–26, 29–32
  CD and ASPD, 76–77
  environment and, 245
  genes affecting, 27–28, 47, 48–49, 80–82
  human inclination toward, and God in Bible, 223–24
  as inborn in individual, 243
  as "integrative element in any healthy society," 161–62
  parenting children at genetic risk for, 245
  predicting, using DNA, 16–17
  psychopathy, 77, 78–79
  questions about genetic association with, 104–5
  removing genes affecting, 160–63
  treating, 72, 225–26
  twin and adoption studies, 80
  See also violence
antisocial personality disorder (ASPD), 76, 77

Index

"Anxiety Is the Dizziness of Freedom" (Chiang), 89–90, 108–9, 267n89
Appelbaum, Paul, 123, 224
apples, real and figurative, 239–42
Arbus, Diane, 264n74
Aristotle, 43
Augustine
  on astrology, 105
  basic facts about, 105–6
  and body as punishment, 106–7, 136, 206
  on Fortuna, 236–37
  one's nature and suffering, 226
  validation of secular power and authority of church and, 58–59
  See also original sin
authoritarianism, 199–200, 281n181
autism, 24
awe, 235–36

BALB/cJ mouse, 70, 81
Barrett, Amy Coney, 153
Bateman, Hallie, 158
Beah, Ishmael, 230–31
Becker, Gary, 182
Beckett, Samuel, 82
"A Behavioral Genetic Investigation of Religiosity and Adolescent Problem Behavior" (Harden), 96
behavioral genetics
  addiction and, 26, 67–68, 80–81, 257n26
  as antidote for blame, 23–26, 29–32
  antisocial behaviors and, 29–32, 72, 80–82, 104–5
  areas and limits of influence, 103–5
  belief in, as determinants, 110–11
  change inherited biology and change behavior, 70
  Christian interpretation of, 133–34
  as compromising free will, 107
  criminal behavior and variant on X chromosome, 27–28
  as genetics of sin, 17
  human behavior as result of free will or of, 88–89
  improving genome, 137
  MAOA gene and violence, 47, 48–49
  "polygenic" and environmental correlations, 82–83
  predicting life outcomes, 16–17
  punishment and, 123–25
  removing genes involved in, 160–63
  scientific conventions for naming genes, 260n47
  sexual orientation and, 25–26
  study of, 14
  use of, in criminal cases, 28–29
behavior(s)
  ability to articulate reasons for, 98, 100
  boundary between bad luck and bad, 59–60
  brain's unconscious preparation for action, 98–100
  callous-unemotionality and, 77
  of child and parenting, 113–17, 168–69
  choice and, 11–12, 208, 219–20, 221, 233, 254n11
  classical conditioning and, 166–67
  environment and, 41, 58, 244, 245
  equating, with identity, 132, 244
  executive self (ego) and, 37
  as expression of soul, 241
  fetal alcohol syndrome and, 112–13, 118, 270n112
  imperfect command over our, as result of divine punishment, 106–7, 206
  individual's feelings about, 102
  moral versus biological, 83–84
  nature/nurture debate, 113–14, 115–17, 118–20, 121–25, 127–28, 242, 243, 244–47

of parents, 172–74
power of, 101–2
punishment and hell as way to insure good, 186
religiosity, 95–96, 102
as responding to potential incentives, 183–85
as result of agency or of genetics, 88–89
self-regulation as scientific term for some sins, 18
sphexish, 220, 288n220
See also antisocial behavior(s); behavioral genetics; original sin; responsibility; sin(s), Christian; violence

Between Vengeance and Forgiveness (Minow), 248
biological cooperation, 192–93
biology, 23–26, 29–32, 157, 161
Blake, Katharine, 30–31
blame
 DNA and, 104
 in Garden of Eden, 109
 genetics as antidote for, 23–26, 29–32
 justification for, 21, 22
 nature/nurture debate and, 122–24, 247
 original sin and, 21, 255n21
 responsibility and, 10, 11–12, 254n11
Bloom, Paul, 197
Bouchard, Tom, 93
Breivik, Anders, 226–29
Brown, Peter, 261n58
Brunner, Hans, 46–47
Brunner syndrome, 46–47, 49–50, 261n52
Buck v. Bell (1924), 148
Burr, Esther Edwards, 178

CADM2 gene, 80–81, 161
Cain, 40
Cain (Nemerov), 74–75, 101–2
callous-unemotionality, 77, 80

carceral system
 American, 132–33, 181–86, 187, 188–89, 215, 222, 281n181
 Norwegian, 229–30
Carlson, Tucker, 281n181
Caspi, Avshalom, 27, 113–14
catharsis, 43–44
Chauvin, Derek, 210–11
Chiang, Ted, 89–90, 108–9, 267n89
child soldiers, 230–32
choice(s)
 ability to make, as perception, 219–20, 221
 as consequences of external factors, 233
 guilt and having, 11–12, 254n11
 as justification for punishment, 184–85, 223
 not taken and guilt, 208
 as responding to potential incentives, 183–85
 See also free will
Chouinard, Yvon, 162–63
Christianity
 American carceral system and, 132–33, 215
 avoiding vice and recovering virtue by transcending body, 72
 behavioral genetics as interpreted in, 133–34
 behavior as expression of soul, 241
 behavior equated with identity in, 132
 binary nature of individual and their fate, 132, 133, 233, 242, 243
 flesh as enemy of morality, 247
 forgiveness for sins, 19
 free will in, 58
 human difference, embodiment, suffering and early, 125–28
 Julian of Eclanum, 127–28
 parenting using corporal punishment and, 175–80
 Pelagius, 20, 23, 127, 225
 psychology and, 73–74
 salvation in, 21

Christianity (cont'd)
  sin in, 19, 20, 22
  treatment for antisocial behavior and, 225–26
  "weakness of the flesh" as God's punishment, 106–7, 206
  white Evangelicals' focus on punishment, 180–81
  See also Augustine; original sin
Christianity Today, 25
"circumstantial luck," 52
The City of God (Augustine), 106
classical conditioning, 166–67
Clinton, Bill, 67
"Clover" (Glück), 161
Collins, Simone, 155
Columbine High School shooting, 44–45
conduct disorder (CD), 76–77
Confessions (Augustine), 105
conscious experiences, subjectivity of, 38
"constitutive luck," 57, 58
cooperation, 192–95, 198
creativity as transcendent, 6
Crime and Human Nature (Wilson and Herrnstein), 183
"Crime and Punishment: An Economic Approach" (Becker), 182
CRISPR technology, 53–54
cross-fostering studies, 49

Dallas, Joe, 25
Dante, 17–18, 208
Dare to Discipline (Dobson), 180–81
Dar-Nimrod, Ilan, 133, 137
Darwin, Charles, 193
Davenport, Charles, 149
Davis, Robert, 118
Dawkins, Richard, 122
Dennett, Daniel, 122
depression, cause of, 24
The Descent of Man, and Selection in Relation to Sex (Darwin), 193

desert, definitions of, 7, 8
Determined: A Science of Life Without Free Will (Sapolsky), 31
determinism, 110–11, 134, 205, 207, 245
de Waal, Frans, 193
Diagnostic and Statistical Manual (*DSM*, American Psychiatric Association), 76, 78
Dick, Danielle, 24
Dillard, Annie, 234
disabilities, individuals with severe, 146–48
DNA. *See* behavioral genetics; genetics
Dobbs v. Jackson (2022), 144, 153
Dobson, James, 180–81
Doleac, Jennifer, 187
domestic abuse, 179
Donovan, Brian, 151
Don't Make Me Count to Three: A Mom's Look at Heart-Oriented Discipline (Plowman), 178
dreams, psychoanalytic view of, 10–11
drug addiction
  as chronic, relapsing illness, 66
  effects on brain of, 66
  genetics and, 26, 67–68, 80, 81, 257n26
  in identical twins, 67
  opioids, 26, 64–65, 257n26
  withdrawal symptoms, 64, 69
Durkheim, Émile, 161–62
Dyer, Chester Dean, 76

Easterbrook, Frank, 153
Eastman Unit prison, 13
Eaves, Lindon, 95–96
Edwards, Jonathan, 19, 178
ego (executive function), 37
Ehrlich, Isaac, 182
Eldred, Julie, 26, 257n26
Ellison, Keith, 210–11
emotional empathy, 167–69
empathy, 135–36
Endgame (Beckett), 82

environment and behavior(s), 41, 58, 82–83, 244, 245
epigenetic inheritance, 127
essentialism and eugenics, 137
eugenics
    abortion for reasons of, 152–54
    concept of "purity," 150
    essentialism and, 137
    Holocaust and, 152
    polygenic screening of embryos, 154–57
    poverty and, 25
    racism and, 150–51
    scientists' promotion of, 149–50
    sterilization, 148
    twin studies and, 91
Eve, 141
    See also Adam and Eve
evolution, 71–72, 162–64, 193, 214, 240
"experience of the numinous," 4, 253n4

Fairbairn, Robert, 116–17
Far from the Tree: Parents, Children, and the Search for Identity (Solomon), 44–45, 47
fears and horror stories, 50–51
The Federalist, 156
Fessler, Daniel, 212
fetal alcohol syndrome, 112–13, 118, 270n112
finger-moving study, 98–100
Fiske, Alan, 201–2
Floyd, George, 210–11
forgiveness
    advantage of, over punishment, 214
    cooperation and, 193
    inviting, 214
    as part of human primal history, 213
    for sins, 19
    with punishment, 214–15
    punishment as necessary for, 200
    as strategy for responding to harm, 214

Forgiveness and Mercy (Hampton), 209–10
Fortuna (Roman goddess), 236–37
FOXP2 gene, 81
The Fragility of Goodness (Nussbaum), 242–43
Frankfurt, Harry, 55
fraternal twins, 79–80
Free Agents: How Evolution Gave Us Free Will (Mitchell), 31–32
free will
    ability to make choices as perception, 219–20, 221
    brain's unconscious preparation for action and, 98–100
    in Christianity, 58
    as compromised by original sin, 107
    focus on current situation and, 108–9, 110
    genetics as compromising, 107
    human behavior as result of genetics or of, 88–89
    as justification for punishment, 223
    "manipulated agents" and, 53–55
    psychotic disorders and, 222–23
    responsibility and, 11–12, 254n11
    of self imputed to others, 53
    sense of powerlessness over uncontrollable forces and, 108, 109
    sins as transgressions and, 19, 20
    Sophocles on, 45–46
Freud, Sigmund, 10–11, 43–44

Galton, Francis, 148–49
Gary, Carleton, 41
Gelman, Susan, 124
gender and CD, 76
Gendron, Payton, 150–51
genetics
    agricultural, 159
    avoiding vice and recovering virtue by changing, 72, 160–61
    "candidate" genes for studying, 120
    CRISPR technology and, 53–54

genetics (cont'd)
  drug addiction and, 67–68
  environment and, 41, 58, 244, 245
  as foe, 104
  human genome, 67, 244
  of identical twins, 57
  individual's constitution and, 57
  as legacy of original sin, 246
  mechanisms to enforce
    cooperation, 192
  as necessary for adaptation and
    evolution, 240
  parenting influence on, 115–16
  predicting life outcomes using,
    16–17
  race and, 151
  reproduction and, 47, 57
  single grain effect, 81–82
  treatments for physical conditions
    caused by, 246
  See also behavioral genetics
Genomic Prediction, 155
Glück, Louise, 140, 161, 246
Gnosticism, 126
God
  Adam and Eve and curse of, 41–42
  human inclination toward
    antisocial behavior and, 223–24
  original sin and belief in good,
    20–21, 126–27, 246
  original sin as retribution of, 106
  salvation only through, 21
  as serpent in Garden, 75
  "weakness of the flesh" as
    punishment of, 106–7, 206
Goddard, Henry, 149–50
*God's Law and Order: The Politics of
  Punishment in Evangelical
  America* (Griffith), 181
Gold, Azgad, 224
Gospel Coalition, 133
Gottesman, Irv, 79–80
*The Great Christian Doctrine of
  Original Sin Defended*
  (Edwards), 178
Gregg v. Georgia (1976), 182
Griffin, Richard, 152–53
Griffith, Aaron, 181

Grisel, Judith, 66–67
guilt
  choices not taken and, 208
  expressions of, 212–13
  having choices and, 11–12, 254n11
  original sin and, 21, 255n21

Hampton, Jean, 209–10, 211, 212
Hare, Robert, 78, 169
Harlow, Henry, 115
Harris, Daniel, 117
Harris, Eric, 44–45
Harris, Robert, 112, 113, 117–18,
  134
Hawley, Noah, 179–80
Heine, Steven, 133, 137
He Jiankui, 53–54
Hereditary (movie), 45
hereditary sin. *See* original sin
"Heredity, Environment, and the
  Question 'How?'" (Anastasi),
  119
Herrnstein, Richard, 183, 184, 185
Hieronymi, Pamela, 206
Hill, Darrell, 75, 84, 85
Hillberry, J. D., 249–50
Himmler, Heinrich, 201
Hofstadter, Douglas, 288n220
Hogben, Lancelot, 119
Holmes, Oliver Wendell, 148
Holocaust and eugenics, 152
Hsu, Steve, 155
human genome, 67, 244
human perceptions, 219, 220–21,
  222–23
Hyde, Luke, 173
hyperactivity and MPAs, 41

identical twins
  compared to fraternal twins, 79–80
  development of schizophrenia by,
    79, 80
  differential treatment by mother
    and, 113–14
  drug addiction in, 67
  genetics of, 57

Index

personalities, inclinations, interests, religious values, and abilities of, 57, 93–95
studies of, 91–95, 113–14, 120–21, 268n93
studies of, reared apart, 79, 80, 92–95, 268n93
in utero development of, 90–91
"The Inclination to Evil and the Punishment of Crime—from the Bible to Behavioral Genetics" (Appelbaum and Gold), 224
The Inferno (Dante), 17–18, 208
insanity defense, 217–18
instrumental conditioning, 169–70
Invictus, Titan, 155

J. Dale Wainwright Unit prison, 13
Jackson, Shirley, 55–57
Jesus, 162
Johnson, Lacey, 209
Julian of Eclanum, 127–28
Jung, Carl, 253n4
Just Mercy (Stevenson), 131

Kallikak, Martin, 150
The Kallikak Family: A Study in the Heredity of Feeble-Mindedness (Goddard), 149–50
Kanner, Leo, 24
Kant, Immanuel, 206
Keats, John, 71
Keller, Evelyn Fox, 119, 245
Klebold family, 44–45
Kleiman, Mark, 187–88
Koch, Fred C., 268n93
Koch Industries, 268n93
Kuhse, Helga, 146–47
Kyle, Chris, 201

Lacey, Nicola, 214–15
Landrigan, Jeffrey, 75–77, 78–79, 84–85, 124, 135
Lanza, Adam, 208
Laughlin, Harry, 149

Law for the Prevention of Offspring with Hereditary Diseases (Nazi Germany), 148
"Law of Effect," 170
Lawrence, D. H., 149
Leshner, Alan, 65–66, 68
life outcomes: predicting, using DNA, 16–17
"The Lottery" (Jackson), 55–57
LSD, 5, 8–10
luck
boundary between bad behavior and bad, 59–60
"circumstantial," 52
"constitutive," 57, 58
moral, 51–52, 55–57, 58
Lykken, David, 119

Machiavellianism, 77–78
machines and nature, 5
Malt, Ulrik Fredrik, 228
"manipulated agents," 52–55
MAOA gene, 47, 48–49, 71
"mark of Cain," 40
Matthew (apostle), 132
McBryde Johnson, Harriet, 147–48
Mele, Al, 52–53, 261n52
"metaphysics of the hangman," 22
methylation, 115–16
mice
bred for biomedical research, 70
injecting directly into brains of, 68–69
used in behavioral neuroscience labs, 61–62, 63
Miller v. Alabama (2011), 185
Milton, John, 74, 89
Minnesota Study of Twins Reared Apart (MISTRA), 92–95, 268n93
minor physical anomalies (MPAs), 40–41
Minow, Martha, 231, 233, 248
The Mirage of a Space Between Nature and Nurture (Keller), 119
miscarriages (in pregnancy), 142–46, 157

Index

Mitchell, Keith, 31–32
Moffitt, Terrie, 113–14
Mohler, Albert, 156–57
morality
  biochemistry and, 72–73
  flesh as enemy of, 247
  nature versus, 23, 54–55, 127, 246–47
  of nonhuman animals, 191–92, 193
  obesity and, 72, 73
  as refuge from fortune, 58
  responsibility and, 205–7, 232
  socialization and, 169
  transformations in, 162
  violations of cooperation and, 194–95
  violence and, 201–2
moral luck, 51–52, 55–57, 58
Mozambique, 231–32

Nagel, Thomas, 38, 88–89
narcissism, 77
nature
  machines and, 5
  miscarriage and, 157
  morality versus, 23, 54–55, 127, 246–47
Nature Neuroscience, 15, 18
nature/nurture debate, 113–14, 115–17, 118–20, 121–25, 127–28, 242, 243, 244–47
Nazis, 147–48, 201
Nemerov, Howard, 74–75, 101–2, 264n74
neural plasticity, 5
Never Enough (Grisel), 66–67
The New Jim Crow: Mass Incarceration in the Age of Colorblindness (Alexander), 186
The New Yorker, 23, 56
The New York Times, 23, 154, 208
The New York Times Magazine, 147
Nietzsche, Friedrich, 11, 22, 99, 199, 254n11, 255n21
nightmares, psychoanalytic view of, 10–11

Norway, 226–30
Nussbaum, Martha, 242–43

obesity
  biochemical treatment for, 72–73
  cause of, 23
  as moral failing, 72, 73
One of Us (Seierstad), 228
opioid addiction, 26, 64–65, 69, 257n26
opioids in brain, 65
Orchid, 155
Oresteia (Aeschylus), 171, 203–4
original sin
  act of Adam and Eve and, 19–20
  belief in God and, 20–21
  blame and, 255n21
  as church's solution to theodicy problem, 20–21, 126–27, 246
  as compromising free will, 107
  fallibility of humans and, 72
  genetics as legacy of, 246
  humans' responsibility for behavior and, 207
  as inheritance of insubordinate flesh of Adam and Eve, 106–7, 206
  primacy and inescapability of, 20–21
  as retribution of God, 106
  salvation and, 128
  sin as part of human nature, 22
  societal effects of, 22, 32
  as type of moral luck, 58
  using corporal punishment as parent and, 178–79
  validation of secular power and authority of church, 58–59, 261n58
orphanin FQ receptor, 65, 68
oxycodone, 65
Ozempic, 72–73

Pagels, Elaine, 58, 126–27, 261n58
pain, 171–72, 179–80
Paradise Lost (Milton), 74, 89

Index

parenting
- abusive, through generations in family, 227
- of children at genetic risk for antisocial behavior, 245
- child's emotional empathy and, 168–69
- current most popular book on, 179
- estrangement within family and, 203
- influence on genetics of, 115–16
- punishment versus reward in, 172–74
- singling one child out for negativity or abuse, 113–17
- teens with elevated testosterone levels, 121
- using corporal punishment, 175–80

parenting-industrial complex, modern American, 42
Paul (apostle), 22, 132
Pearl, Michael, 178–79
Pelagius, 20, 23, 127, 225
Pereboom, Derk, 204
Peretti, Frank, 73
personality traits, "dark triad" of, 77–78
Phillips, Carl, 213
Phillips, Emilia, 141
Piaget, Jean, 99
Pickard, Hanna, 187–88, 214–15
Pindar, 242, 243
Pioneer Fund, 93
Plowman, Ginger, 178
Poetics (Aristotle), 43
Pollan, Michael, 4, 253n4
positive reinforcement, 168–70, 171–74
Powell, Marcia, 7–8, 22, 222, 254n8
This Present Darkness (Peretti), 73
Preterm-Cleveland v. McCloud (United States Court of Appeals for the Sixth Circuit, 2021), 152–53
Prometheus (Scott movie), 43
Promising Young Woman (movie), 195

Prozac, 5
psychedelic drugs, 5, 8–10
psychology
- "Boulder model" of graduate study of, 87–88
- as un-Christian, 73–74

psychopathy, 77, 78–79
psychosis, 219
psychotic disorders, 222–23
psychotic-type experiences, 218–19
"public goods games," 194
the Punisher (Marvel character), 200–202
punishment
- arbitrary, 55–57
- as asserting equality, 210–11
- authoritarian punitiveness as aspect of, 200
- being capable of great evil as, 136
- choice as justification for, 184–85, 223
- community service as, 211
- death penalty as, 182
- divine, 106
- effective alternatives to, 187–88
- to establish and maintain dominance relationships, 211–12, 213
- with forgiveness, 214–15
- forgiveness's advantage over, 214
- genetic inheritance and, 29–32
- as its own justification, 184
- justification for, 21, 22
- nature/nurture debate and, 123–25
- as necessary for forgiveness, 200
- Norwegian compared to American attitudes, 229
- as opposite of reward, 170, 171–74
- parenting using corporal, 175–80
- psychotic disorders and, 222
- retribution and, 199
- to satisfy feelings of rage and powerlessness, 189
- sensitivity to pain and, 171
- as strategy for responding to harm, 214

punishment (*cont'd*)
  "weakness of the flesh" as Adam's and Eve's, 106–7, 206
  white Evangelical Christians' focus on, 180–81
  See also carceral system
Purdue Pharma, 65
"Putting It Together" (Hillberry), 249–50

"The Queerness of Eve" (Phillips), 141

race/racism
  eugenics and, 150–51
  distinct from genetics, 151
  incarceration rate and, 186
Rai, Tage, 201–2
rape, 195–96, 210
The Reckonings (Johnson), 209
religiousness, 95–96, 102
reproduction
  abortion, 143–44, 145, 152–57
  in apple industry, 239–40
  humans as "breeding true," 242, 243
  inheritance of genes during, 47, 57
  miscarriages, 142–46, 157
  selective breeding, 152–57, 158–64
  sex-linked mutations and, 47
  sexual, 206
  sterilization, 148
  uniqueness in each instance of, 47
responsibility
  blame and, 10, 11–12, 254n11
  determinism and, 205
  as ineluctable, 207
  insanity defense and, 218
  of "manipulated agents," 54–55
  of men with undiagnosed Brunner syndrome, 261n52
  as moral issue, 205–7, 232
  moral versus legal, 218
  of nonrational "choosers," 221

  original sin and, 207
  as social practice, 232
  "Responsibility and the Limits of Evil" (Watson), 134–35
retribution
  activity characteristic of pleasure as neural signature of, 196–99
  carceral state and, 188
  as deeply rooted and deeply dangerous, 205
  as humbling, 210–11
  original sin as, of God, 106
  power and, 211–12, 213
  punishment and, 199

Sanger, Margaret, 149
Sapolsky, Robert, 31, 204
Satel, Sally, 257n26
Scalia, Antonin, 185
schizophrenia in twins, 79–80
Science, 27, 65–66
scientific journals, 15–16, 17
  See also specific journals
"Scientists Can Design 'Better' Babies. Should They?" (*The New York Times*), 154
Scott, Ridley, 43
Seierstad, Åsne, 228
self-consciousness, requirement of, 38–39
self-regulation, as scientific term for some sins, 18
sensorimotor stage of development, 98–100
serotonin, 5
sex, 106–7, 195–96, 206, 210
sexual orientation and behavioral genetics, 25–26
Shakespeare, William, 118
shame, 38–39, 212–13
Should the Baby Live? The Problem of Handicapped Infants (Singer and Kuhse), 146–47
Shriver, Lionel, 44
Sierra Leone, 230–31
Singer, Peter, 146–47

Index

sin(s), Christian
    behavioral genetics as genetics of, 17
    gluttony, 72, 73
    as part of human nature, 22
    scientific terms and, 18
    as transgressions, 19, 20
    types of, 19
    See also original sin
60 Minutes II, 84
Skinner, B. F., 171–73, 183–84, 186
Slater, Lauren, 5, 226
Smith, Angela, 230–32
Socrates, 162
solar eclipses, 233–35, 237
Solomon, Andrew, 44–45, 47
Sophocles, 45–46
sphexishness, 220, 288n220
Spufford, Francis, 19
"stark fiction," 45–46
sterilization, 148
Stevenson, Bryan, 131, 185
Stewart, Potter, 182
"Stocking Strangler," 41
Strawson, Galen, 254n11
Strawson, Peter, 192, 205, 206–7
The Study of Human Heredity: Methods of Collecting, Charting, and Analyzing Data (Davenport and Goddard), 149
sympathy, 134–35

Tabarrok, Alex, 185
Teeth Are Not for Biting (Verdick), 167
The Tempest (Shakespeare), 118
Texas Heartbeat Act (2021), 143–44
Texas Monthly, 14
Thinking About Crime (Wilson), 183
Thorndike, Edward, 170
Three Identical Strangers (documentary), 93
Tillich, Paul, 20

toe webbing, 40
To Train Up a Child (Pearl), 178–79
transcendence and creativity, 6
twins, 79–80, 95, 105
    See also identical twins

The Uninnocent: Notes on Violence and Mercy, (Blake), 30–31

victims and genetics, 25
The View from Nowhere (Nagel), 88–89
violence
    CD and ASPD, 77
    child soldiers, 231–32
    family described by Brunner, 46–47
    FOXP2 gene and, 81
    MAOA gene and, 47, 48–49
    mass, 44–45, 226–29, 248
    moral outrage and, 201–2
    MPAs' correlation with, 40, 41
    truth commissions and war-crimes trials and closure after, 248
    variables in probability of, 248–49
    variant on X chromosome and, 27–28
Virtuous Violence (Fiske and Rai), 201–2
volitional self. *See* free will

Walden Two (Skinner), 172–73
Waldron, Mary, 92
Waldroup, Bradley, 26–27, 28
Waldroup, Penny, 26–27
The Washington Post, 95–96
Watson, Gary, 134–35
Watts, Alan, 109
Weber, Max, 95, 132
We Need to Talk About Kevin (Shriver), 44
When Brute Force Fails (Kleiman), 187

When Should Law Forgive? (Minow), 231, 233
Whitfield, Ruth, 150
The Wild Iris (Glück), 140
Williams, Bernard, 45–46
Wilson, James Q., 182–83, 184, 185
Winnicott, Donald, 115
Wolf, Susan, 160

The Women of Trachis (Sophocles), 45–46
women scientists, 68–69
  See also specific individuals
Wynn, Karen, 197

Yates, Andrea, 216–18, 220, 221–22, 225

ABOUT THE AUTHOR

KATHRYN PAIGE HARDEN is a professor in the department of psychology at the University of Texas at Austin, where she leads the Developmental Behavior Genetics lab. Harden received her PhD in clinical psychology from the University of Virginia and completed her clinical internship at McLean Hospital/Harvard Medical School. She has been honored by the American Psychological Association for her distinguished scientific contributions to the study of genetics and human individual differences.

ABOUT THE TYPE

This book was set in Caledonia, a typeface designed in 1939 by W. A. Dwiggins (1880–1956) for the Mergenthaler Linotype Company. Its name is the ancient Roman term for Scotland, because the face was intended to have a Scottish-Roman flavor. Caledonia is considered to be a well-proportioned, businesslike face with little contrast between its thick and thin lines.2w3qeewqsaa